VOLUME 6: HOW TO ANALYZE RELIABILITY DATA

Wayne Nelson

American Society for Quality Control
611 East Wisconsin Avenue
Milwaukee, Wisconsin 53202

ISBN 0-87389-018-3

The ASQC Basic References in Quality Control: Statistical Techniques (John A. Cornell, Ph.D. and Samuel S. Shapiro, Ph.D., Editors) is a continuing literature project of ASQC's Statistics Division. Its aims are to survey topics in statistical quality control, show what sorts of problems they can solve, and tell where further techniques of the type discussed can be located in practical (usable) form, and to provide the quality practitioner on the everyday firing line of a manufacturing plant with specific ready-to-use tools for conducting statistical analysis in the quality investigative process.

Suggestions on this booklet or for new booklets are welcome, and these should be sent to either editor.

Volumes Published

How To Analyze Data With Simple Plots (W. Nelson)
How To Perform Continuous Sampling (CSP) (K. S. Stephens)
How To Test Normality And Other Distributional Assumptions (S. S. Shapiro)
How To Perform Skip-Lot And Chain Sampling (K. S. Stephens)
How To Run Mixture Experiments For Product Quality (J. A. Cornell)
How To Analyze Reliability Data (W. Nelson)
How And When To Perform Buyesian Acceptance Sampling
How To Apply Response Surface Methodology
How To Use Regression Analysis In Quality Control
How To Plan An Accelerated Life Test

In order to purchase volumes, write to: American Society for Quality Control, 611 East Wisconsin Avenue Milwaukee, Wisconsin 53202, U.S.A.

FOREWORD

The ASQC Basic References in Quality Control: Statistical Techniques is a literature project of the Statistics Division of ASQC. The series' Review Board consists of Saul Blumenthal, Joseph W. Foster, Alan J. Gross, Gerald J. Hahn, Norman L. Johnson, H. Alan Lasater, Edward A. Sylvestre and Harrison M. Wadsworth, Jr., supplemented (for the current volume) by Nancy R. Mann.

This volume deals with the techniques of analyzing reliability data. It presents modern methods for extracting from life test and field data the information needed to make sound decisions. The presentation is on an elementary level so those who have had only a basic course in statistics can apply the techniques. There are numerous references to more extensive discussions of the techniques and to computer programs which can supply the needed calculations. The booklet begins with a discussion of the basic concepts and theory for product life distributions and then describes graphical and analytical techniques used to analyze life data.

Dr. Nelson is a consulting statistician with the General Electric Company. He also consults privately on and teaches engineering and scientific applications of statistical methods for other companies. In 1981 he received General Electric's *Dushman Award* in recognition of his outstanding contributions to research and applications in product life data analysis and accelerated testing. He is also an Adjunct Professor both at Union College and Rensselaer Polytechnic Institute where he teaches courses on the theory and application of statistics. He is a Fellow of the American Statistical Association and a member of the American Society for Quality Control. Much of the material presented in this booklet was developed in conjunction with his work. He is one of the recognized leaders in this subject and recently published a book, *Applied Life Data Analysis*.

Samuel S. Shapiro
Florida International University
Miami, Florida
January 1983

ACKNOWLEDGMENT

The author is grateful to Mr. William Cleveland of the Management Problems Analyses Program, Manufacturing Education, General Electric Management Development Institute, Crotonville, New York, for supporting the development of the material for this booklet and to John Wiley and Sons, New York for permission to use material from the author's book *Applied Life Data Analysis*. The author thanks Dr. Samuel S. Shapiro and the Editorial Review Board for their considerable help on this volume.

TABLE OF CONTENTS

HOW TO ANALYZE RELIABILITY DATA

by
Wayne Nelson

General Electric Co. Corporate Research & Development
Schenectady, NY 12345

CHAPTER I LIFE DISTRIBUTIONS AND CONCEPTS

Almost every major company yearly spends millions of dollars on product reliability. Much management and engineering effort goes into evaluating risks and liabilities, predicting warranty costs, evaluating replacement policies, assessing design changes, identifying causes of failure, and comparing alternate designs, vendors, materials, manufacturing methods, and the like. Major decisions are based on product life data, often from a few units. This volume presents modern methods for extracting from life test and field data the information needed to make sound decisions. Such methods are successfully used on a great variety of products by many who have just a working knowledge of basic statistics from a first course.

This chapter presents basic concepts and theory for product life distributions, used as models for the life of products, materials, people, television programs, and many other things. The commonly used exponential, normal, lognormal, and Weibull distributions are presented below and are used to analyze data graphically in Chapter II and numerically in Chapter III. Chapter I also presents the Poisson and binomial distributions, which are probability models for the observed numbers of failures or defectives. For further detail, consult the books referenced in Chapter IV, which surveys topics not covered in this booklet.

1. BASIC CONCEPTS AND THE EXPONENTIAL DISTRIBUTION

The cumulative distribution function $F(y)$ for a continuous distribution represents the population fraction failing by age y. Any such $F(y)$ has the mathematical properties:

a) it is a continuous function for all y,

b) $\lim_{y\to-\infty} F(y) = 0$ and $\lim_{y\to\infty} F(y) = 1$, and

c) $F(y) \leqslant F(y')$ for all $y < y'$.

The exponential cumulative distribution function for the population fraction failing by age y is

$$F(y) = 1 - e^{-y/\theta}, \; y \geqslant 0.$$

$\theta > 0$ is the mean time to failure, θ is in the same measurement units as y, for example, hours, months, cycles, etc. Figure I.1A shows this cumulative distribution function. In terms of the "failure rate" $\lambda = 1/\theta$,

$$F(y) = 1 - e^{-\lambda y}, \; y \geqslant 0.$$

Engine fan example. The exponential distribution with a mean of $\theta = 28{,}700$ hours was used to describe the hours to failure of a fan on diesel engines. The failure rate is $\lambda = 1/28{,}700 = 34.9$ failures per million hours. For the engine fans, the population fraction failing on an 8,000 hour warranty is $F(8{,}000) = 1 - \exp(-8{,}000/28{,}700) = 0.24$.

The reliability function $R(y)$ for a life distribution is the probability of survival beyond age y, namely,

$$R(y) \equiv 1 - F(y).$$

The exponential reliability function is

$$R(y) = e^{-y/\theta}, \; y \geqslant 0.$$

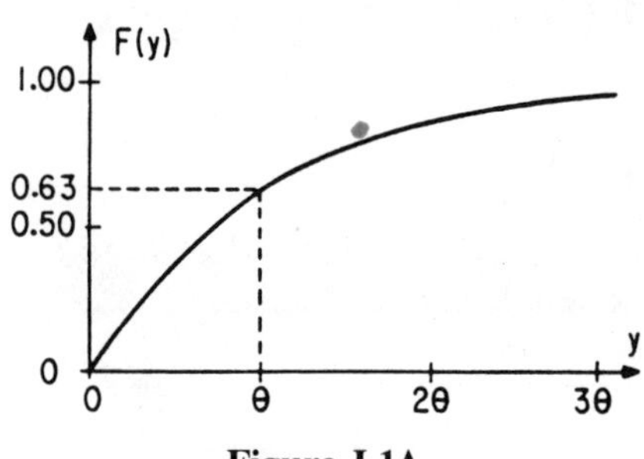

Figure I.1A
Exponential Cumulative Distribution

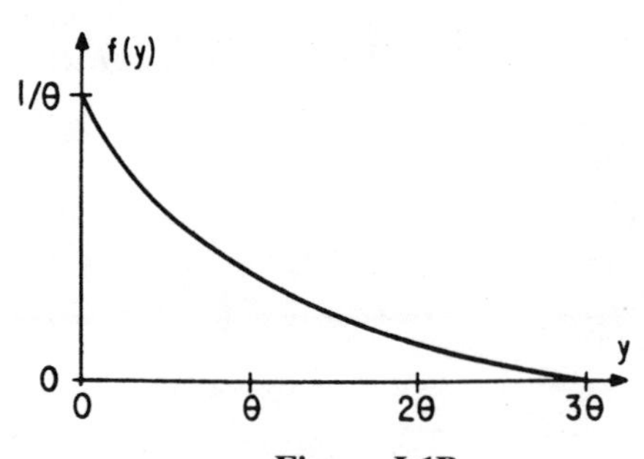

Figure I.1B
Exponential Probability Density

For the engine fans, reliability for 8,000 hours is $R(8{,}000) = \exp(-8{,}000/28{,}700) = 0.76$. That is, 76% of such fans survive warranty.

The 100Pth percentile of a distribution $F(\ )$ is the age y_P by which a proportion P of the population fails. It is the solution of

$$P = F(y_P).$$

In life data work, one often wants to know low percentiles such as the 1% and 10% points, which correspond to early failure. The 50% point is called the *median* and is commonly used as a 'typical' life.

The 100Pth exponential percentile is

$$y_P = -\theta \ln(1-P).$$

For example, the mean θ is roughly the 63rd percentile of the exponential distribution. For the diesel engine fans, median life is $y_{.50} = -28{,}700 \ln(1-0.50) = 19{,}900$ hours. The 1st percentile is $y_{.01} = -28{,}700 \ln(1-0.01) = 288$ hours.

The probability density of a cumulative distribution function is

$$f(y) \equiv \frac{dF\,(y)}{dy}.$$

It corresponds to a histogram of the population life times.

The exponential probability density is

$$f(y) = (1/\theta)\ e^{-y/\theta},\ y \geqslant 0.$$

Figure I.1B depicts this probability density. Also,

$$f(y) = \lambda\ e^{-\lambda y},\ y \geqslant 0.$$

The mean μ of a distribution with probability density $f(y)$ is

$$\mu \equiv \int_{-\infty}^{\infty} y\, f(y)\, dy.$$

The integral runs over all possible outcomes y. The mean is also called the *average* or *expected life*. It corresponds to the arithmetic average of the lives of all units in a population. It is used as still another 'typical' life.

The mean of the exponential distribution is

$$\mu = \int_{0}^{\infty} y\ (1/\theta)\ e^{-y/\theta}\, dy = \theta.$$

This shows why θ is called the mean time to failure (MTTF). Also, $\mu = 1/\lambda$. For the diesel engine fans, the mean life is $\mu = 28{,}700$ hours. Some repairable equipments have exponentially distributed time *between* failures, particularly after most components have been replaced a few times. Then θ is called the mean time between failures (MTBF).

The hazard function $h(y)$ of a distribution is defined as

$$h(y) \equiv f(y)/[1-F(y)] = f(y)/R(y).$$

It is the *(instantaneous) failure rate* at age y. That is, in the short time Δ from y to $y+\Delta$, a proportion $\Delta \cdot h(y)$ of the population that reached age y fails. $h(y)$ is a measure of proneness to failure as a function of age. It is also called the *hazard rate* and the *force of mortality*. In many applications, one wants to know whether the failure rate of a product increases or decreases with product age.

The exponential hazard function is

$$h(y) = [(1/\theta)\ e^{-y/\theta}]/e^{-y/\theta} = 1/\theta,\ y \geqslant 0.$$

Figure I.1C shows this constant hazard function. Also, $h(y) = \lambda$, $y \geqslant 0$. Only the exponential distribution has a constant failure rate, a key characteristic. That is, for this distribution only, an old unit and a new unit have the same chance of failing over a future time interval Δ. For example, engine fans of any age will fail at a constant rate of $h(y) = 34.8$ failures per million hours.

A decreasing hazard function during the early life of a product is said to correspond to *infant mortality*. Figure I.2 shows this near time zero. Such a failure rate often indicates that the product is poorly designed or suffers from manufacturing defects. Some products, such as some semiconductor devices, have a decreasing failure rate over their observed life.

An increasing hazard function during later life of a product is said to correspond to *wear-out* failure. This often indicates that failures are due to the product wearing out. Figure I.2 shows this feature in the later part of the curve. Many products have an increasing failure rate over the entire range of life.

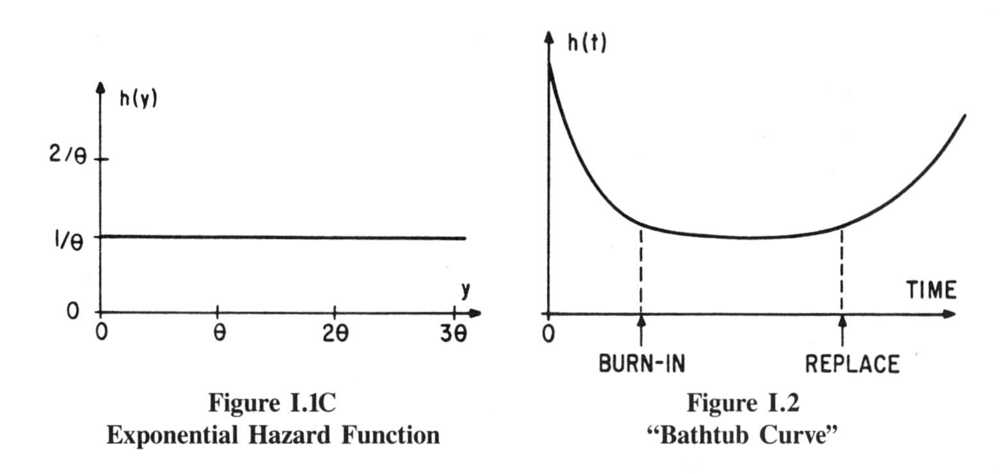

Figure I.1C
Exponential Hazard Function

Figure I.2
"Bathtub Curve"

The bathtub curve. A few products show a decreasing failure rate in the early life and an increasing failure rate in later life. Figure I.2 shows such a hazard function, called a "bathtub curve." Some products, such as high-reliability capacitors and semiconductor devices, are subjected to a burn-in. This weeds out early failures before units are put into service. Also, units are removed from service before wearout starts. Thus units are in service only in the low failure rate portion of their life. This increases their reliability in service. Jensen and Petersen (1982) comprehensively treat planning and analysis of burn-in procedures, including the economics.

Distributions for Special Situations

Distributions with failure at time zero. A fraction of a population may already be failed at time zero. Consumers may encounter a product that does not work when purchased. The model for this consists of the proportion p failed at time zero and a continuous life distribution for the rest. Such a cumulative distribution appears in Figure I.3A. The sample proportion failed at time zero is used to estimate p, and the failure times in the remainder of the sample are used to estimate the continuous distribution.

Distributions with eternal survivors. Some units may never fail. This applies to 1) the time to death from a disease when some individuals are immune, 2) the time to redemption of trading stamps (some stamps are lost and never redeemed), 3) the time to product failure from a particular defect when some units lack that defect, and 4) time to warranty claim on a product whose warranty applies only to original owners, some of which sell the product before failure. Figure I.3B depicts this situation.

Mixtures of distributions. A population may consist of two or more subpopulations. Figure I.3C depicts this situation. Units from different production periods may have different life distributions due to differences in design, raw materials, environment, etc. It is often important to identify such a situation and the production period, customer, environment, etc., that has poor units. Then suitable action may be taken on that portion of the population. A mixture should be distinguished from competing failure modes, described in Section 7.

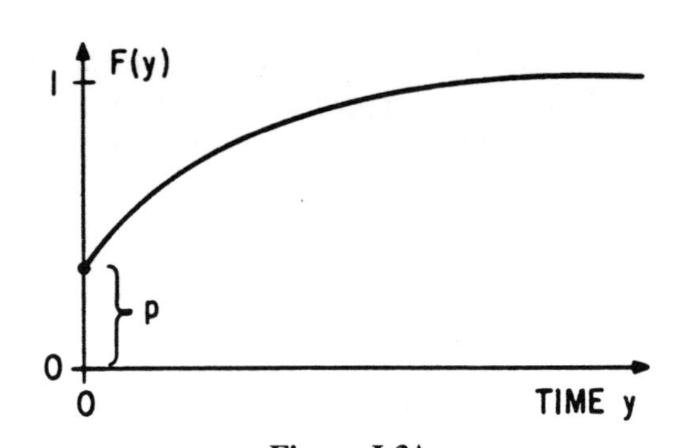

Figure I.3A
A Cumulative Distribution with a Fraction Failed at Time Zero

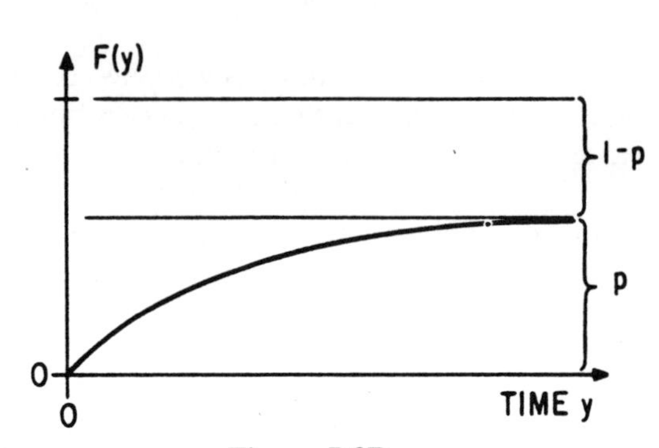

Figure I.3B
A Cumulative Distribution with Eternal Survivors

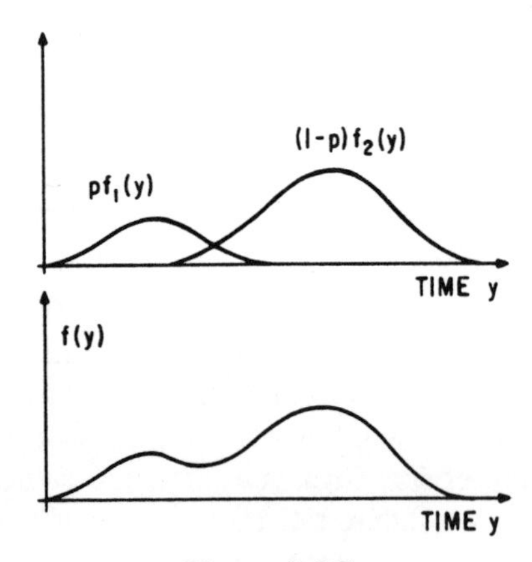

Figure I.3C
A Mixture of Distributions

2. NORMAL DISTRIBUTION

This section presents the normal distribution. Its hazard function increases. Thus it may describe products with wear-out failure.

The normal probability density is

$$f(y) = (2\pi\sigma^2)^{-1/2}\exp[-(y-\mu)^2/(2\sigma^2)], \quad -\infty < y < \infty.$$

μ is the population mean and may have any value. σ is the population standard deviation and must be positive. μ and σ are in the same measurement units as y, for example, hours, months, cycles, etc. Figure I.4A depicts this probability density, which is symmetric about the mean μ. The figure shows that μ is the distribution median and σ determines the spread.

The range of y is from $-\infty$ to $+\infty$. Life must, of course, be positive. Thus the distribution fraction below zero must be small for this distribution to be a satisfactory approximation in practice.

The normal cumulative distribution function for the population fraction failing by age y is

$$F(y) = \int_{-\infty}^{y} (2\pi\sigma^2)^{-1/2} \exp[-(x-\mu)^2/(2\sigma^2)]\, dx, \quad -\infty < y < \infty.$$

Figure I.4B depicts this function. This can be expressed in terms of the standard normal cumulative distribution function $\Phi(\ )$ as

$$F(y) = \Phi[(y-\mu)/\sigma], \quad -\infty < y < \infty.$$

Many tables of $\Phi(z)$ give values only for $z \geqslant 0$. One then uses $\Phi(-z) = 1 - \Phi(z)$.

Transformer example. A normal life distribution with $\mu = 6250$ hours and $\sigma = 2600$ hours was used to represent life of a transformer. The fraction of the distribution with negative life times is $F(0) = \Phi[0-6250)/2600] = \Phi(-2.52) = 0.0059$. This small fraction is ignored hereafter.

The 100P-th normal percentile is

$$y_P = \mu + z_P\sigma\ ;$$

here z_P is the 100P-th standard normal percentile and is tabled below. The *median* (50th percentile) of the normal distribution is $y_{0.50} = \mu$, since $z_{0.50} = 0$.

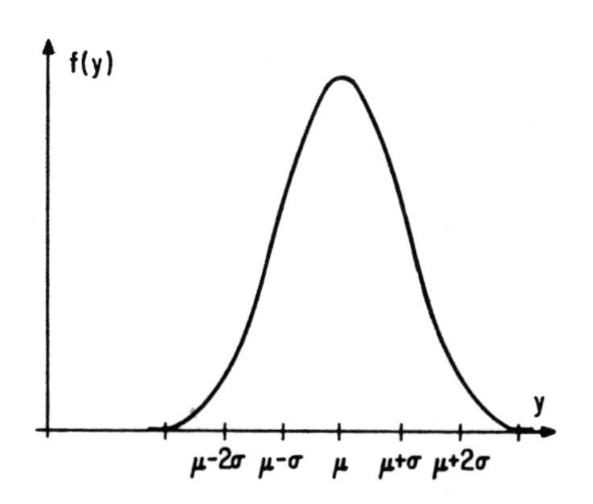

Figure I.4A
Normal Probability Density

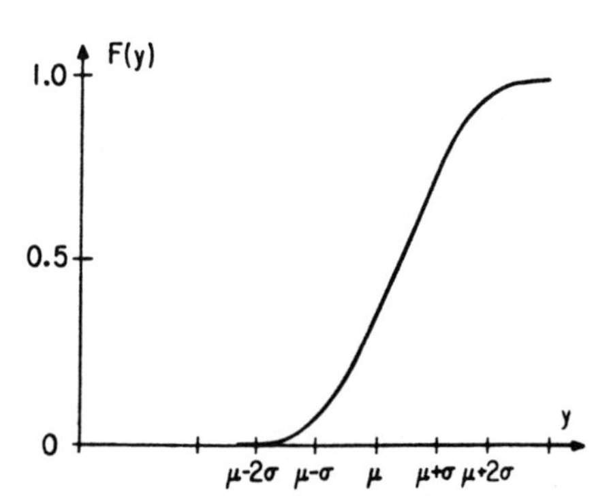

Figure I.4B
Normal Cumulative Distribution

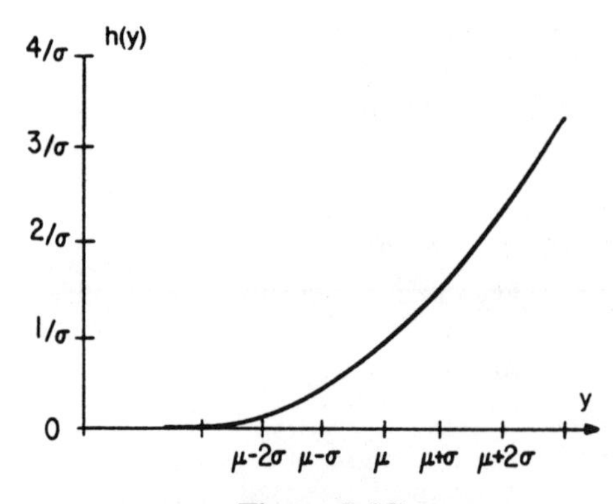

Figure I.4C
Normal Hazard Function

Some standard percentiles are:

100P%:	0.1	1	2.5	5	10	50	90	97.5	99
z_P:	−3.090	−2.326	−1.960	−1.645	−1.282	0	1.282	1.960	2.326

Median transformer life is $y_{0.50} = 6250$ hours, and the 10th percentile is $y_{0.10} = 6250+(-1.282)2600 = 2920$ hours.

The normal hazard function appears in Figure I.4C, which shows that the normal distribution has an *increasing failure rate* (wearout) with age.

A key question was: does transformer failure rate increase with age? If so, older units should be replaced first. The increasing failure rate of the normal distribution indicates that older units are more failure prone.

3. LOGNORMAL DISTRIBUTION

The lognormal distribution is used for certain types of life data, for example, metal fatigue and electrical insulation life. The lognormal and normal distributions are related, a fact used to analyze lognormal data with methods for normal data.

The lognormal probability density is

$$f(y) = \{0.4343/[(2\pi)^{1/2}y\sigma]\} \exp\{ - [\log(y)-\mu]^2/(2\sigma^2)\},\ y > 0.$$

μ is called the *log mean* and may have any value; it is the mean of the *log* of life — not of life. σ is called the *log standard deviation* and must be positive; it is the standard deviation of the *log* of life — not of life. μ and σ are not 'times' like y; instead they are unitless pure numbers. $0.4343 \cong 1/\ln(10)$. Here log( ) denotes the common (base 10) logarithm. Some authors use the natural (base e) logarithm, denoted by ln( ). Figure I.5A shows probability densities, which have a variety of shapes. The value of σ determines the shape of the distribution, and the value of μ determines the 50% point and the spread.

The lognormal cumulative distribution function for the population fraction failing by age y is

$$F(y) = \Phi\ \{[\log(y)-\mu]/\sigma\},\ y > 0.$$

Figure I.5B shows lognormal cumulative distribution functions.

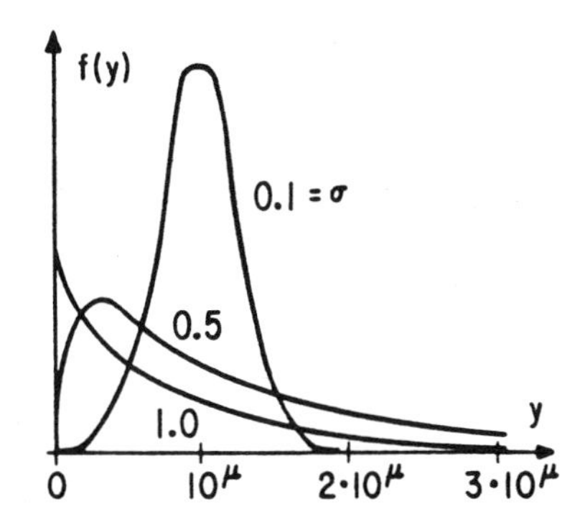

Figure I.5A
Lognormal Probability Densities

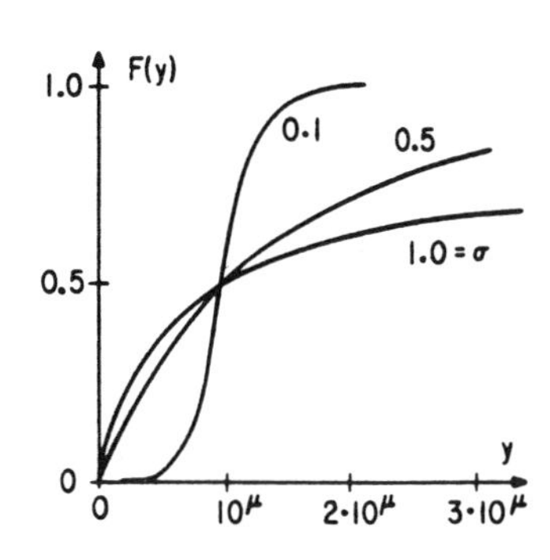

Figure I.5B
Lognormal Cumulative Distributions

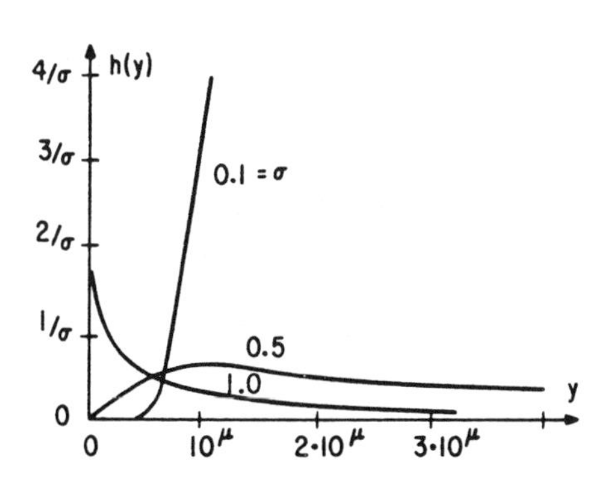

Figure I.5C
Lognormal Hazard Functions

Locomotive control example. The life (in thousand miles) of an electronic control for locomotives was approximated by a lognormal distribution where $\mu = 2.236$ and $\sigma = 0.320$. The population fraction failing on an 80 thousand mile warranty is $F(80) = \Phi[(\log(80)-2.236)/0.320] = \Phi[-1.04] = 0.15$. This percentage was too high, and the control was redesigned.

The 100P-th lognormal percentile is

$$y_P = \text{antilog}[\mu+z_P\sigma];$$

here z_P is the 100P-th standard normal percentile. The *median* (50th percentile) is $y_{0.50} = \text{antilog}\ [\mu]$.

For the locomotive control, $y_{0.50} = \text{antilog}[2.236] = 172$ thousand miles, regarded as a typical life. The 1% life is $y_{0.01} = \text{antilog}[2.236+(-2.326)0.320] = 31$ thousand miles.

Lognormal hazard functions appear in Figure I.5C. For $\sigma \simeq 0.5$, $h(y)$ is essentially constant. For $\sigma \leq 0.2$, $h(y)$ increases and is much like that of a normal distribution. For $\sigma \geq 0.8$, $h(y)$ decreases. This flexibility makes the lognormal distribution popular and suitable for many products. The lognormal hazard function has a property seldom seen in products. It is zero at time zero, increases to a maximum, and then decreases to zero with increasing age. However, over most of its range, the lognormal distribution does fit life data on many products.

For the locomotive control, $\sigma = 0.320$. So the behavior of $h(y)$ is midway between the increasing and roughly constant hazard functions in Figure I.5C.

The relationship between the lognormal and normal distributions helps one understand the lognormal distribution in terms of the simpler normal distribution. The (base 10) log of a variable with a lognormal distribution with parameters μ and σ has a normal distribution with mean μ and standard deviation σ. Thus the analysis methods for normal data can be used for the logarithms of lognormal data.

4. WEIBULL DISTRIBUTION

The Weibull distribution is often used for product life, because it describes increasing and decreasing failure rates. It may be suitable for a "weakest link" product; i.e., the product consists of many parts with comparable life distributions and the product fails with the first part failure. For example, the life of a capacitor is determined by the shortest-lived portion of its dielectric.

The Weibull probability density function is

$$f(y) = (\beta/\alpha^{\beta})\ y^{\beta-1} \exp[-(y/\alpha)^{\beta}],\ y > 0.$$

The *shape parameter* β and the *scale parameter* α are positive. α is called the *characteristic life*, as it is always 63.2th percentile. α has the same units as y, for example, hours, months, cycles, etc. β is a unitless pure number. The Weibull probability densities in Figure I.6A show that β determines the shape of the distribution and α determines the spread. For $\beta = 1$, the Weibull distribution is the exponential distribution. For much life data, the Weibull distribution is more suitable than the exponential, normal, and lognormal distributions.

The Weibull cumulative distribution function for the population fraction failing by age y is

$$F(y) = 1 - \exp[-(y/\alpha)^{\beta}],\ y > 0.$$

Figure I.6B shows Weibull cumulative distribution functions.

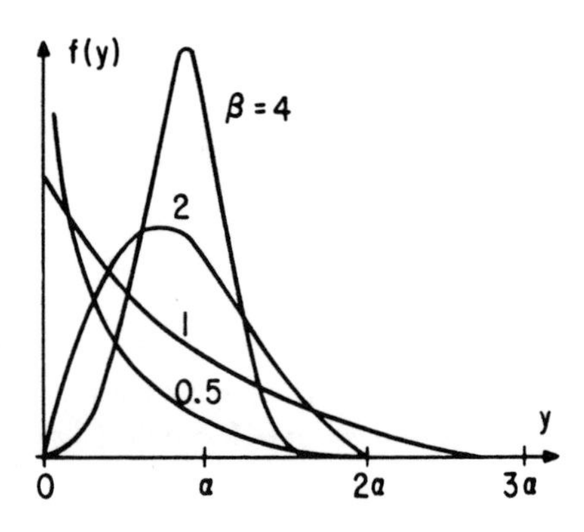

Figure I.6A
Weibull Probability Densities

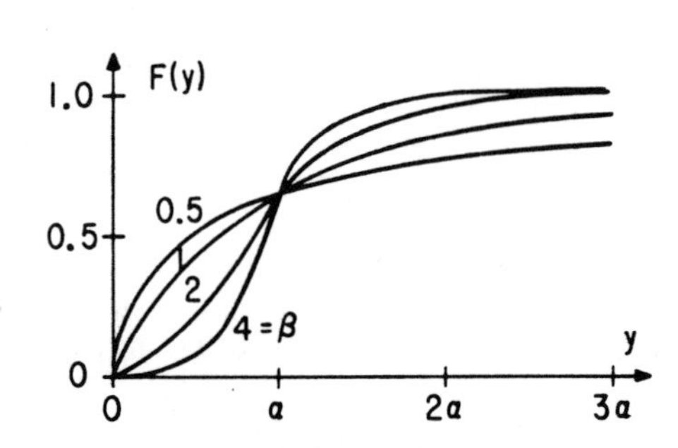

Figure I.6B
Weibull Cumulative Distributions

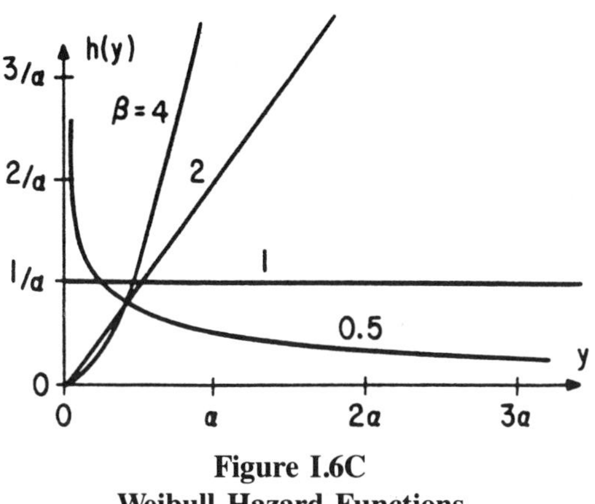

Figure I.6C
Weibull Hazard Functions

Winding example. The life of generator field windings was approximated by a Weibull distribution where $\alpha = 13$ years and $\beta = 2$. The population fraction of windings failing on a two-year warranty is $F(2.0) = 1 - \exp[-(2.0/13)^2] = 0.023$ or 2.3%.

The Weibull reliability function for the population fraction surviving beyond age y is

$$R(y) = \exp[-(y/\alpha)^\beta], \; y > 0.$$

For the windings, the population reliability for two years is $R(2.0) = \exp[-(2.0/13)^2] = 0.977$ or 97.7%.

The 100P-th Weibull percentile is

$$y_P = \alpha[-\ln(1-P)]^{1/\beta} \; ;$$

here ln( ) is the natural logarithm. For example, $y_{0.632} \cong \alpha$ for any Weibull distribution. This may be seen in Figure I.6B.

For the windings, $y_{0.632} = 13[-\ln(1-0.632)]^{1/2} = 13$ years, the characteristic life. The 10th percentile is $y_{0.10} = 13[-\ln(1-0.10)]^{1/2} = 4.2$ years.

The Weibull hazard function is

$$h(y) = (\beta/\alpha)\,(y/\alpha)^{\beta-1}, \; y > 0.$$

Figure I.6C shows Weibull hazard functions. A power function of time, $h(y)$ increases for $\beta > 1$ and decreases for $\beta < 1$. For $\beta = 1$ (the exponential distribution), the failure rate is constant. With increasing or decreasing failure rates, the Weibull distribution flexibly describes product life.

For the windings, $\beta = 2$, and their failure rate increases with age, wear-out behavior. This tells utilities that preventive replacement of old windings will avoid costly failures in service.

5. POISSON DISTRIBUTION

The Poisson distribution is used for the number of occurrences of some event within some observed time, area, volume, etc. For example, it has been used to describe the yearly number of soldiers of a Prussian regiment kicked to death by horses, the number of flaws in a length of wire or computer tape, the number of failures of a repairable product over a certain period, and many other phenomena. It is appropriate if 1) the occurrences occur independently of each other over time (area, volume, etc.), 2) the chance of an occurrence is the same for each point in time (area, volume, etc.), and 3) the potential number of occurrences is unlimited.

The Poisson probability of y occurrences is

$$f(y) = (1/y!)\,(\lambda t)^y \, e^{-\lambda t}, \; y = 0,1,2,\ldots \; .$$

Here t is the amount of exposure or observation; it may be a time, length, area, volume, etc. For example, for a power line, t is the product of length and time in thousand-ft-years. The *occurrence rate* λ must be

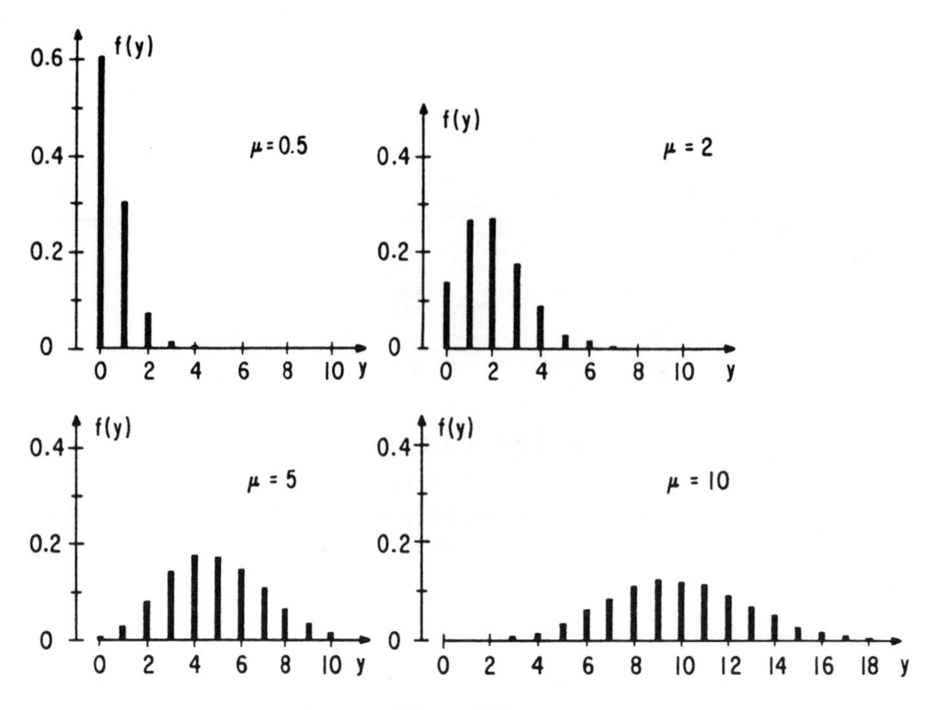

Figure I.7A
Poisson Probability Functions

positive; it is the expected number of occurrences per unit time, length, area, volume, etc. Figure I.7A depicts Poisson probability functions.

Power line. For a power line, the yearly number of failures is assumed to have a Poisson distribution with $\lambda = 0.0256$ failures per year per thousand feet. For $t = 515.8$ thousand feet of line, the probability of no failures in a year is $f(0) = (1/0!)(0.0256\times515.8)^0 \exp(-0.0256\times515.8) = \exp(-13.2) = 1.8\times10^{-6}$. So the possibility of no failures is negligible.

The Poisson cumulative distribution function for the probability of y or fewer occurrences is

$$F(y) = \sum_{i=0}^{y} (1/i!)(\lambda t)^i \, e^{-\lambda t}.$$

The Thorndike chart in Figure I.7B provides $F(y)$ as follows. Enter the chart on the horizontal axis at the value $\mu = \lambda t$. Go up to the curve labeled y. Then go horizontally to the vertical scale to read $F(y)$. For the power line, $\lambda t = 13.2$, and the probability of 15 or fewer failures is $F(15) = 0.75$ from the chart. $F(y)$ is tabulated in most textbooks.

The Poisson mean of the number Y of occurrences is

$$\mu = \lambda t.$$

For the power line, the expected (mean) number of failures in a year is $\mu = 0.0256(515.8) = 13.2$ failures, useful in maintenance planning.

The Poisson standard deviation of the number Y of occurrences is

$$\sigma(Y) = (\lambda t)^{1/2}.$$

For the power line, $\sigma(Y) = [13.2]^{1/2} = 3.63$ failures in a year.

A normal approximation to the Poisson $F(y)$ is

$$F(y) \cong \Phi\,[\,(y+0.5 - \lambda t) \,/\, (\lambda t)^{1/2}];$$

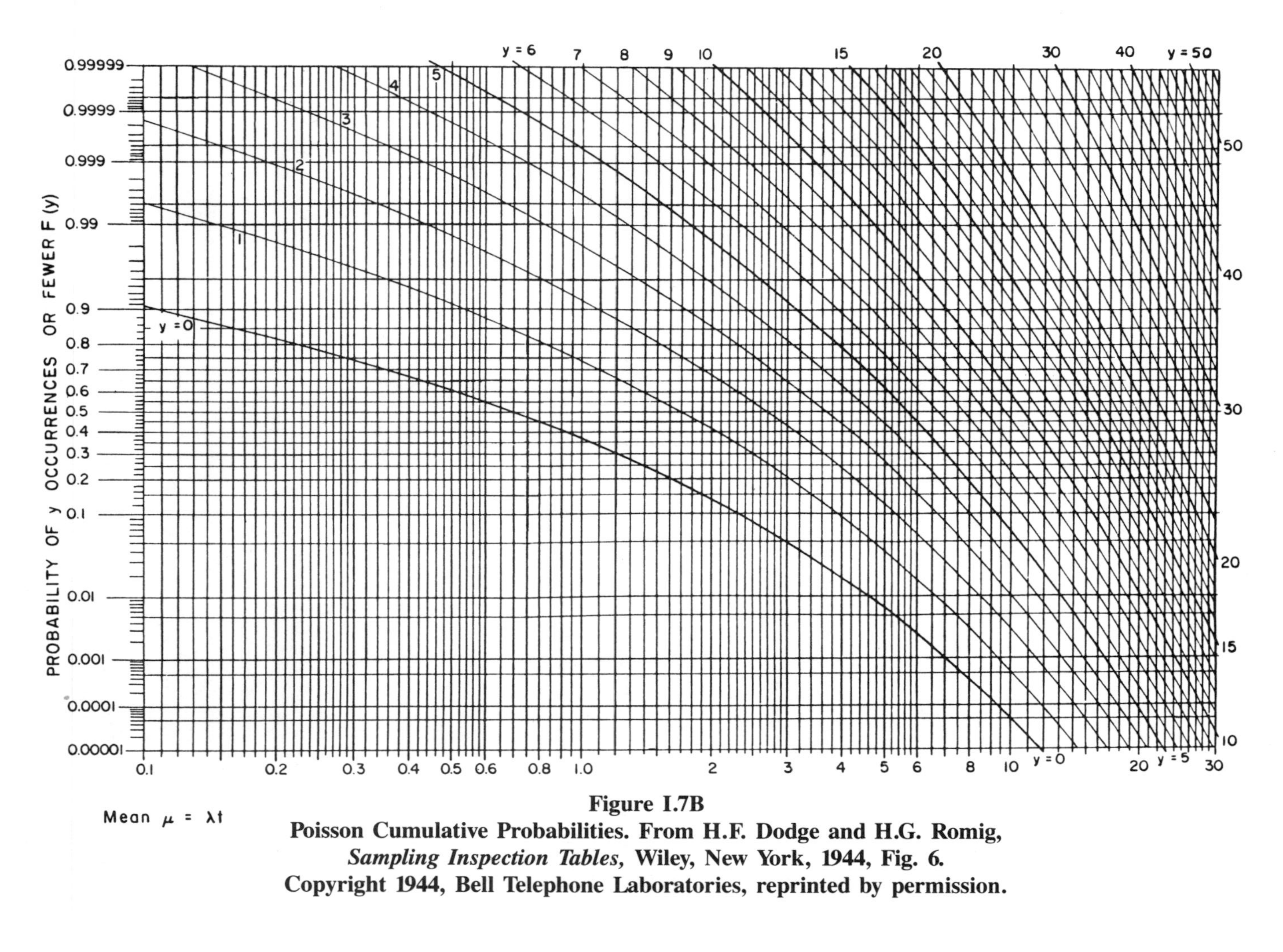

Figure I.7B

Poisson Cumulative Probabilities. From H.F. Dodge and H.G. Romig, *Sampling Inspection Tables*, Wiley, New York, 1944, Fig. 6.

here Φ [ ] is the standard normal cumulative distribution function. This approximation is satisfactory for most practical purposes if $\lambda t \geqslant 10$.

For the power line, the approximate probability of 15 or fewer failures in a year is $F(15) \cong \Phi[(15+0.5-13.2)/3.63] = \Phi[0.63] = 0.74$. The exact probability is 0.75.

Demonstration testing commonly involves the Poisson distribution. Repairable hardware "demonstrates" its reliability if units run a specified total time t with y or fewer failures. Units that fail are repaired and kept on test. A manufacturer designs the hardware to achieve a λ that assures passing the test with a desired high probability $100(1-\alpha)\%$. The hardware can fail the test with $100\alpha\%$ probability, called the *producer's risk.*

Electronic System Example. An electronic system was required to run $t = 10{,}000$ hours with $y = 2$ or fewer failures. For the electronic system, the producer's risk was to be 10%. To obtain the desired design λ, one must find λt such that the Poisson probability $F_{\lambda t}(y) = 1-\alpha$. To do this, enter Figure I.7B on the vertical axis at $1-\alpha$, go horizontally to the curve for y or fewer failures, and then go down to the horizontal axis to read the appropriate μ value. Then the desired design failure rate is $\lambda = \mu/t$. For the electronic system, $1-\alpha = 1-0.10 = 0.90$, $\mu = 1.15$, and $\lambda = 1.15/10{,}000 = 0.115$ failures per thousand hours.

Relationship of Poisson and exponential distributions. For a repairable product, suppose that times between failures are statistically independent and have an exponential distribution with failure rate λ. Then the *number* of failures in a total running time t over any number of units has a Poisson distribution with mean λt.

6. BINOMIAL DISTRIBUTION

The binomial distribution is used as a model for the number of sample units that fall in a specified category. For example, it is used for the number of defective units in samples from shipments and production, the number of units that fail on warranty, and the number of one-shot devices (used once) that work properly.

Its assumptions are (a) each sample unit has the same chance p of being in the category and (b) the outcomes of the n sample units are statistically independent.

The binomial probability of getting y category units in a sample of n units is

$$f(y) = \frac{n!}{y!\,(n-y)!}\,p^y\,(1-p)^{n-y},\ y = 0,1,2,\ldots,n;$$

p is the population proportion in the category ($0 \leqslant p \leqslant 1$). Figure I.8 depicts binomial probability functions.

In reliability work, if the category is "failure" of a device, the proportion p is the *failure probability*, sometimes incorrectly called the failure rate. The proportion p is expressed as a percentage and differs from the Poisson failure rate λ, which has the dimensions of failures per unit time. If the category is "successful operation" of a device, the proportion p is called the *reliability* of the device.

A locomotive control under development was assumed to fail on warranty with probability $p = 0.156$. $n = 96$ such controls were field tested, and $y = 15$ failures occurred on warranty. The probability is $f(15) = 96![15!(96-15)!]^{-1}\,(0.156)^{15}\,(1-0.156)^{96-15} = 0.111$.

The binomial cumulative distribution function for the probability of y or fewer sample items in the category is

$$F(y) = \sum_{i=0}^{y} \frac{n!}{i!\,(n-i)!}\,p^i\,(1-p)^{n-i},\ y = 0,1,2,\ldots,n.$$

F(y) is widely tabulated. For example, the probability of 15 or fewer warranty failures of the 96 controls is $F(15) = 0.571$ from a binomial table.

A normal approximation is

$$F(y) \cong \Phi\,\{(y+0.5-np)/[np(1-p)]^{1/2}\}\ ;$$

here Φ[ ] is the standard normal cumulative distribution function. This approximation is usually satisfactory if $np \geqslant 10$ and $n(1-p) \geqslant 10$.

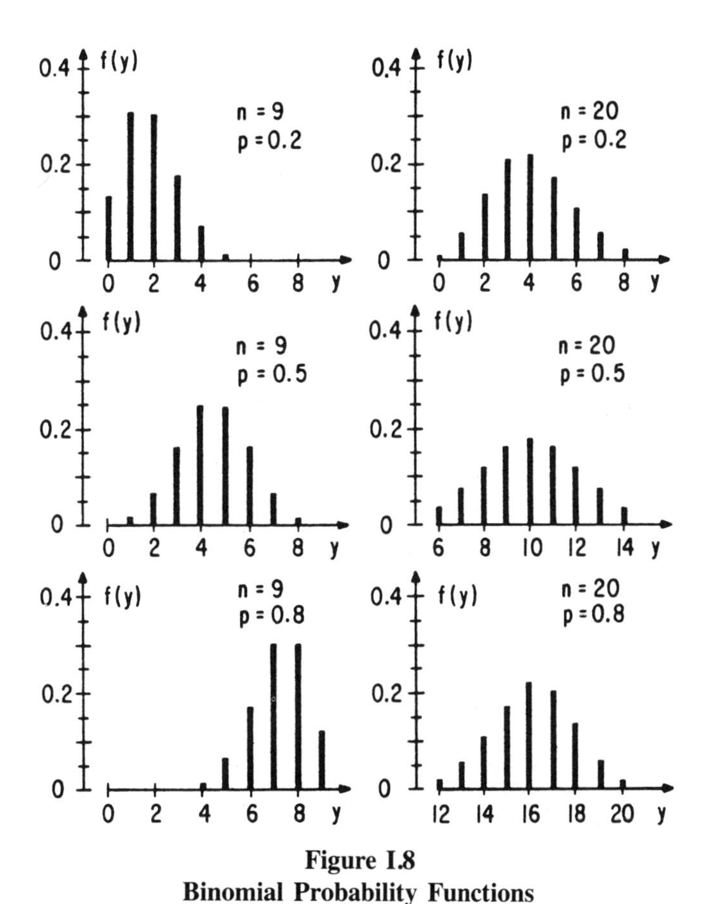

Figure I.8
Binomial Probability Functions

For example, for the controls, $F(15) \cong \Phi\ \{(15+0.5-96\times0.156)/[96\times0.156\ (1-0.156)]^{1/2}\} = 0.556$. Similarly, the approximate probability of 15 failures is $f(15) = F(15)-F(14) \cong 0.556 - \Phi\ \{(14+0.5-96\times0.156)/[96\times0.156\ (1-0.156)]^{1/2}\} = 0.112$ (0.111 exact). Here $96(0.156) = 15 > 10$ and $96(1-0.156) = 81 > 10$.

The mean of the number Y of sample items in the category is

$$\mu = np.$$

This is the number n of sample units times the population proportion p in the category. For example, the mean number of failures in samples of 96 locomotive controls is $\mu = 96\times0.156 = 15.0$ failures.

Acceptance sampling plans based on the binomial distribution appear in MIL-STD-105D and in quality control books, for example, Grant and Leavenworth (1972) and Schilling (1982).

An acceptance sampling plan specifies the number n of sample units and the acceptable number y of defective units in the sample. If the sample contains y or fewer defective units the product passes; otherwise, it fails. A plan had $n = 20$ and $y = 1$. If the product has a proportion defective of $p = 0.01$, the chance it passes inspection is $F(1) = f(0)+f(1) = \dfrac{20!}{0!(20-0)!}\ 0.01^{0}\ 0.99^{20} + \dfrac{20!}{1!(20-1)!}\ 0.01^{1}\ 0.99^{19} = 0.983$, which could be read from a binomial table. The chance of passing as a function of p is called the *Operating Characteristic* (OC) *Curve* of the plan (n,y). The OC curve for $n = 20$ and $y = 1$ appears in Figure I.9.

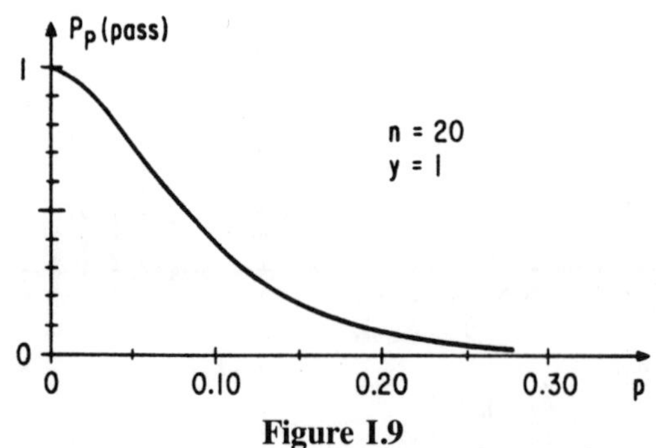

Figure I.9
OC Curve of Acceptance Sampling Plan

7. SERIES SYSTEMS AND MULTIPLE CAUSES OF FAILURE

Many products fail from more than one cause. For example, any part in an appliance may fail and cause the appliance to fail. Also, humans may die from accidents, various diseases, etc. The series-system model represents the relationship between the product life distribution and those of its parts. Chapter II presents graphical analyses of data with a number of causes of failure.

This section presents the series-system model, the product rule for reliability, the addition law for failure rates, and the resulting distribution when some failure modes are eliminated.

Series systems and the product rule. Suppose that a product has a potential time to failure from each of M causes (also called competing risks or failure modes). Such a product is called a *series system* if its life is the smallest of those M potential times to failure. This is, the first part failure produces system failure.

Let $R(y)$ denote the system reliability function and let $R_1(y),\ldots,R_M(y)$ denote the reliability functions of the M causes (each in the absence of all other causes). It is assumed that the M potential times to failure of a system are statistically *independent*. Such systems are said to have *independent competing risks* or to be *series systems* with independent causes of failure. For such systems, it can be shown that

$$R(y) = R_1(y)R_2(y) \ldots R_M(y) .$$

This key result is the *product rule* for reliability of series systems (with independent components).

Three-way bulb. By engineering definition, a three-way light bulb fails if either filament fails. Filament 1 (2) has a normal life distribution with a mean of 1500 (1200) hours and a standard deviation of 300 (240) hours. Filament reliability functions are depicted as straight lines on normal probability paper in Figure I.10. The life distribution of such bulbs was needed, in particular, the median life. Filament lives are assumed independent; so the bulb reliability is $R(y) = \{1-\Phi[(y-1500)/300]\} \times \{1-\Phi[(y-1200)/240]\}$; here $\Phi[\ ]$ is the standard normal cumulative distribution. For example, $R(1200) = \{1-\Phi[(1200-1500)/300]\} \times \{1-\Phi[(1200-1200)/240]\} = 0.421$. $R(y)$ is plotted in Figure I.10 and is not quite a straight line (not a normal distribution). The median life is obtained by solving $R(y_{0.50}) = 0.50$ to get $y_{0.50} = 1160$ hours; this also can be obtained from the plot.

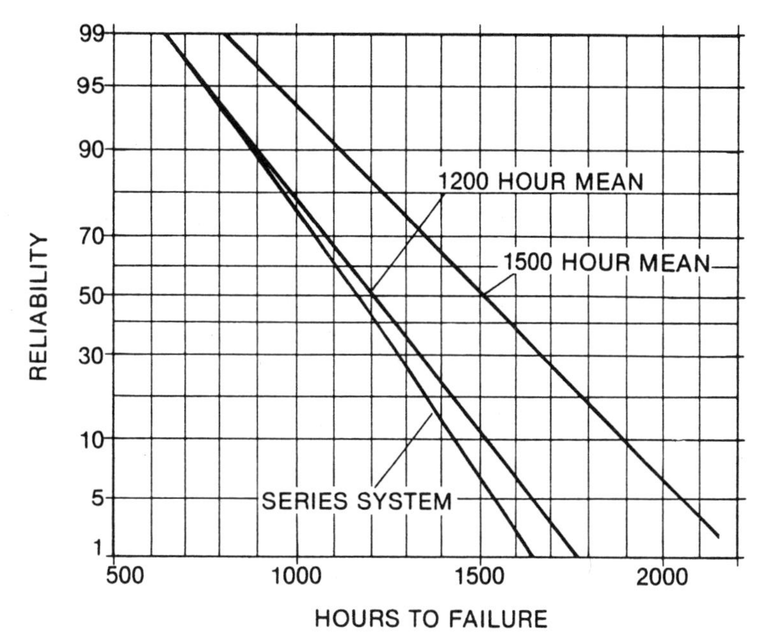

Figure I.10
Reliability Function of a Bulb with Two Filaments

Addition law for failure rates. Denote the system hazard function by $h(y)$ and those for the failure causes by $h_1(y),\ldots, h_M(y)$. Then it can be shown that

$$h(y) = h_1(y) + h_2(y) + \ldots + h_M(y) \;;$$

this is called the *addition law for failure rates* for *independent* failure modes (or competing risks). This law is depicted in Figure I.11, which shows the hazard functions of the two components of a series system and the system hazard function.

A pronounced increase in the failure rate of a product may occur at some age. This may indicate that a new failure cause with an increasing failure rate is becoming dominant at that age, as in Figure I.11.

Exponential causes. Suppose that M independent causes have *exponential* life distributions with failure rates $\lambda_1,\ldots,\lambda_M$. Then series systems consisting of such components have an exponential life distribution with a constant failure rate

$$\lambda = \lambda_1 + \ldots + \lambda_M.$$

This simple relationship is often *incorrectly* used for reliability analysis of systems with components that do not have constant failure rates. Then the previous equation is correct.

Freight train. A high-priority freight train was required to have its three locomotives all complete a one day run. If a locomotive failed, the railroad had to pay a large penalty. The railroad needed to know the reliability of such trains. Time to failure for such locomotives has an exponential distribution with $\lambda_0 = 0.023$ failures per day. A series system, the train has an exponential distribution of time to delay with $\lambda = 0.023 + 0.023 + 0.023 = 0.069$ delays per day. Train reliability on a one day run is $R(1) = \exp(-0.069\times1) = 0.933$.

Elimination of failure modes. Often it is important to know how elimination of some causes of failure will improve product life. Suppose that cause 1 is eliminated (this may be a collection of causes). Then $R_1(y) = 1$, $h_1(y) = 0$, and the life distribution for the remaining causes has

$$R^*(y) = R_2(y) \ldots R_M(y),\; h^*(y) = h_2(y) + \ldots + h_M(y)\;.$$

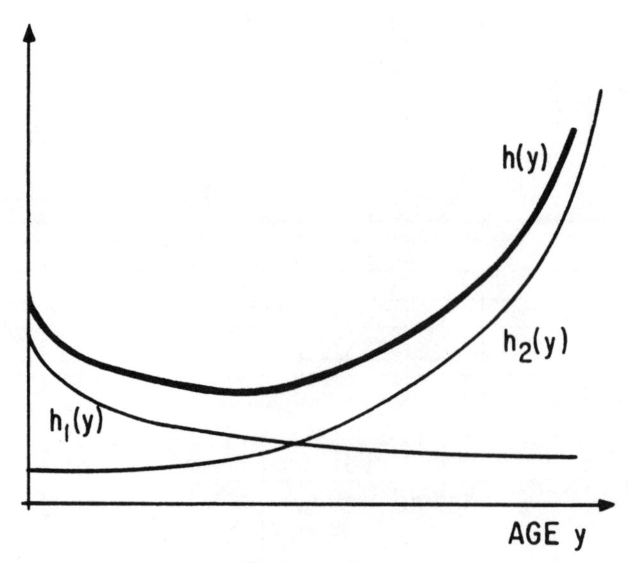

Figure I.11
Hazard Functions of a Series System and Its Components

If the 1500-hour filament were replaced by one with essentially unlimited life, the bulb would have the life distribution of the 1200-hour filament.

Series systems with dependence. Some series-system products contain parts with statistically *dependent* lifetimes. For example, adjoining segments of a cable may have positively correlated lives, that is, have similar lives. Models for dependent part lives are complicated; David and Moeschberger (1979) and Block and Savits (1981) comprehensively survey multivariate distributions with dependence.

CHAPTER II GRAPHICAL ANALYSES OF LIFE DATA

This chapter presents graphical analyses of life data. Section 1 discusses needed background for data analysis. Section 2 provides graphical estimates of the life distribution, percentage failing on warranty, etc. Section 3 provides an estimate of the life distribution that would result if certain failure modes were eliminated. Section 4 provides an estimate of the life distribution of a single failure mode. For further detail, consult Nelson (1982) and books referenced in Section 12 of Chapter IV.

1. BACKGROUND

Background for data analysis is briefly presented here. The topics are population and sample, valid data, failure and exposure, types of data, and needed information.

Population and sample. A life distribution describes some *population.* A manufacturer of fluorescent bulbs is concerned with the life distribution of future production of a certain bulb — an essentially infinite population. A generator manufacturer is concerned with the life of a population of units in service and to be manufactured in coming years. We analyze data from a *random sample* (set of units) to get numerical information on the population distribution.

Valid data. It is assumed that the sample is randomly taken from the population of interest. A sample from another population or a subset of the population may give misleading information. For example, failure data from appliances on a service contract overestimate failure rates for appliances not on contract. Also, lab test data may greatly differ from field data. In practice, one must often use questionable data. Then one must assess how well such data represent the population of interest and how much one can rely on the information.

Failure and exposure. Failure must be precisely defined in practice, especially in dealings between producers and consumers. For many products, failure is catastrophic, and it is clear when failure occurs. For some products, performance slowly degrades, and there is no clear end of life. Engineering can then define that a failure occurs when performance degrades below a specified value. One may analyze data for each of several definitions of failure. Engineering must decide whether to use calendar time, operating hours, or some other measure of exposure, such as the number of start-ups, miles traveled, or cycles of operation.

Types of data. The proper analysis of data depends on the type of data. The following paragraphs describe the common types of life data from tests and actual service.

Life data are *complete* if the time to failure of each sample unit is known. Figure II.1A depicts a complete sample. For much life data, the exact failure times of some units are unknown, and there is only partial information on their failure times. Examples follow.

When units are unfailed, their failure times are beyond their present survival times. Such data are said to be *censored.* If all unfailed units have a common survival time and all failure times are earlier, the data are called *singly censored.* Such data result when units are started on test together and the data are analyzed before all units fail. Figure II.1B depicts such a sample. If differing survival times are intermixed with the failure times, the data are called *multiply censored.* Field data are usually multiply censored, since units go into service at different times and have different survival times when the data are analyzed. Figure II.1C depicts such a sample.

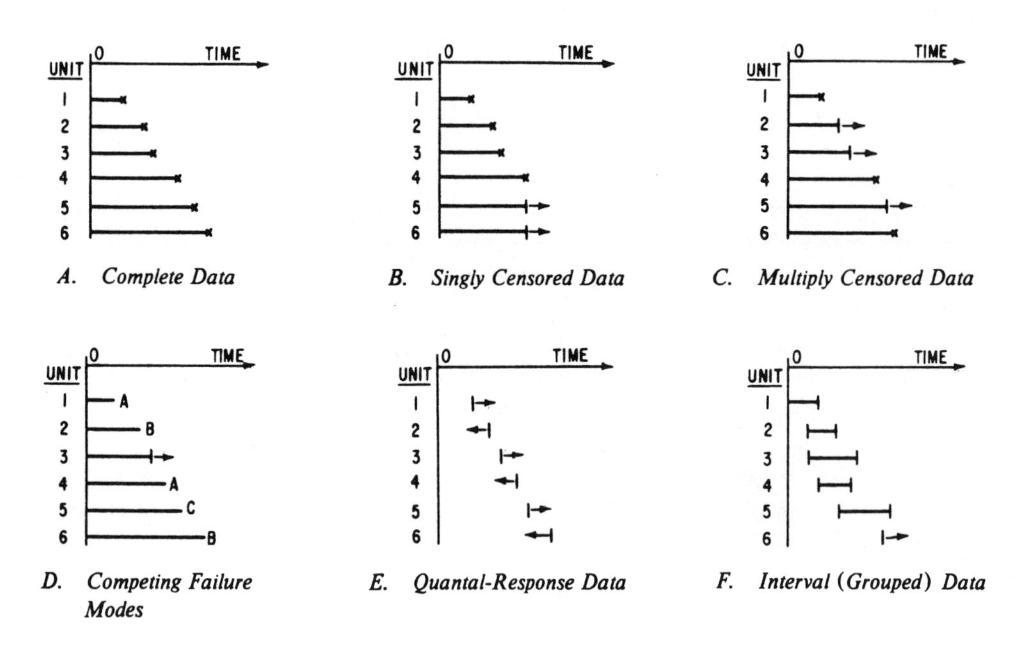

Figure II.1
Types of Data (failed ×, survived ⊢→, failed earlier ←⊣)

Data have *competing failure modes* when sample units fail from different causes. Figure II.1D depicts such a sample with causes A, B, and C.

Sometimes one knows only whether a unit failed before or after it was inspected once, revealing, for example, a cracked part in a turbine. Different units usually have different ages when inspected. Such data are *quantal-response* data. Figure II.1E depicts such a sample.

When units are inspected for failure more than once, one knows only that a unit failed in an interval between inspections. So-called *interval* data are depicted in Figure II.1F. Such data can also contain units unfailed at their last inspection.

Needed information. One must specify what *numerical* information on the population is needed to draw practical conclusions and make decisions. Such population information may be the proportion failing on warranty, the mean or median life, and the life distribution that would result if redesign eliminated certain failure modes. Of course, data analysis provides no decisions. It only provides numerical information for people who make decisions. If such people have difficulty specifying the numerical information they need, they should imagine that all the population data are available and then decide what values calculated from the data would be useful. Data analysis provides estimates of such population values from small samples.

2. HAZARD PLOTTING OF LIFE DATA

Data plots are used for display and interpretation of data because they are simple and effective, as described by Nelson (1979). Hazard plots are widely used to analyze field and life test data on products consisting of electronic and mechanical parts and ranging from small electrical appliances through heavy industrial equipment. This section presents hazard plots to estimate a life distribution from multiply censored life data. Such plots do not apply to failures found on inspection, since the failure occurred earlier at an unknown time; analyses for inspection data appear in Chapter III.

Appliance component. Data that illustrate hazard plotting appear in Table II.1, which shows the cycles (number of times used) to failure of a component of a small appliance in a development program. Each survival time has a + to indicate that the failure time of the unfailed component is beyond. Failure times are unmarked. Engineering wanted an estimate of the percentage failing on warranty (500 cycles) and an estimate of median life.

Steps to Make a Hazard Plot

1. Order the n times from smallest to largest as shown in Table II.1 without regard to which are survival of failure times. Label the times with reverse ranks; that is, label the first time with n, the second with $n-1,\ldots,$ and the n'th with 1.

2. Calculate a hazard value for each *failure* as $100/k$, where k is its reverse rank, as shown in Table II.1. For example, the failure at 145 cycles has reverse rank 46, and its hazard value is 100/46 = 2.2%. Hazard values are tabulated in Table II.2.

3. Calculate the cumulative hazard value for each *failure* as the sum of its hazard value and the cumulative hazard value of the preceding failure. For example, for the failure at 145 cycles, the cumulative hazard value of 6.0 is the hazard value 2.2 plus the previous cumulative hazard value 3.8. Cumulative hazard values appear in Table II.1. Cumulative hazard values have no physical meaning and may exceed 100%.

4. Choose a hazard paper. There are hazard papers* for the exponential, Weibull, extreme value, normal, and lognormal distributions. The distribution is often chosen from engineering knowledge of the product.

5. On the vertical axis of the hazard paper, mark a time scale that brackets the data. For the component data, normal hazard paper was chosen, and marked from 0 to 1200 cycles as shown in Figure II.2.

6. On the paper, plot each failure time vertically against its cumulative hazard value on the horizontal axis as shown in Figure II.2. Survival times are *not* plotted; hazard and cumulative hazard values are not calculated for them. However, the survival times do determine the proper plotting positions of the failure times through the reverse ranks. STATPAC of Nelson and others (1978) and other computer programs do such calculations and make such plots.

7. If the plot of failure times is roughly straight, the theoretical distribution fits the data. By eye, fit a straight line through the data points. Also, one can just use the plotted points without a line.

The line estimates the cumulative percentage failing (read from the horizontal probability scale at the top of the grid) as a function of age. The straight line, as explained below, yields information on the life distribution. If the plot is curved, plot the data on another hazard paper. If no hazard paper yields a straight enough plot, draw a smooth curve through the plotted data. Then, as described below, use the curve in the same way as a straight line to estimate percentiles and failure probabilities.

The basic assumption. Hazard plotting is valid if the life distribution of units censored at a given age is the same as the life distribution of units that run beyond that age. For example, this assumption is not satisfied if units are removed from service when they look like they are about to fail.

---

* Offered in the Catalog of TEAM, Box 25, Tamworth, N.H. 03886, (603) 323-8843.

Table II.1
Hazard Calculations for the Appliance Component

Cycles	Reverse Rank k	Hazard 100/k	Cum. Hazard
45+	54		
47	53	1.9	1.8
73	52	1.9	3.8
136+	51		
136+	50		
136+	49		
136+	48		
136+	47		
145	46	2.2	6.0
190+	45		
190+	44		
281+	43		
311	42	2.4	8.4
417+	41		
485+	40		
485+	39		
490	38	2.6	11.0
569+	37		
571+	36		
571	35	2.9	13.9
575	34	2.9	16.8
608+	33		
608+	32		
608+	31		
608+	30		
608	29	3.4	20.2
608+	28		
608+	27		
608+	26		
608+	25		
608+	24		
608+	23		
608	22	4.6	24.8
608+	21		
608+	20		
630	19	5.3	30.1
670	18	5.6	35.7
670	17	5.9	41.6
731+	16		
838	15	6.7	48.3
964	14	7.1	55.4
964	13	7.7	63.1
1164+	12		
1164+	11		
1164+	10		
1164+	9		
1164+	8		
1164+	7		
1164+	6		
1198+	5		
1198	4	25.0	88.1
1300+	3		
1300+	2		
1300+	1		

Table II.2
Hazard Values 100/k for k = 1 to 200

k	100/k	k	100/k	k	100/k	k	100/k
1	100.00	51	1.96	101	.99	151	.66
2	50.00	52	1.92	102	.98	152	.66
3	33.33	53	1.89	103	.97	153	.65
4	25.00	54	1.85	104	.96	154	.65
5	20.00	55	1.82	105	.95	155	.65
6	16.67	56	1.79	106	.94	156	.64
7	14.29	57	1.75	107	.93	157	.64
8	12.50	58	1.72	108	.93	158	.63
9	11.11	59	1.69	109	.92	159	.63
10	10.00	60	1.67	110	.91	160	.62
11	9.09	61	1.64	111	.90	161	.62
12	8.33	62	1.61	112	.89	162	.62
13	7.69	63	1.59	113	.88	163	.61
14	7.14	64	1.56	114	.88	164	.61
15	6.67	65	1.54	115	.87	165	.61
16	6.25	66	1.52	116	.86	166	.60
17	5.88	67	1.49	117	.85	167	.60
18	5.56	68	1.47	118	.85	168	.60
19	5.26	69	1.45	119	.84	169	.59
20	5.00	70	1.43	120	.83	170	.59
21	4.76	71	1.41	121	.83	171	.58
22	4.55	72	1.39	122	.82	172	.58
23	4.35	73	1.37	123	.81	173	.58
24	4.17	74	1.35	124	.81	174	.57
25	4.00	75	1.33	125	.80	175	.57
26	3.85	76	1.32	126	.79	176	.57
27	3.70	77	1.30	127	.79	177	.56
28	3.57	78	1.28	128	.78	178	.56
29	3.45	79	1.27	129	.78	179	.56
30	3.33	80	1.25	130	.77	180	.56
31	3.23	81	1.23	131	.76	181	.55
32	3.12	82	1.22	132	.76	182	.55
33	3.03	83	1.20	133	.75	183	.55
34	2.94	84	1.19	134	.75	184	.54
35	2.86	85	1.18	135	.74	185	.54
36	2.78	86	1.16	136	.74	186	.54
37	2.70	87	1.15	137	.73	187	.53
38	2.63	88	1.14	138	.72	188	.53
39	2.56	89	1.12	139	.72	189	.53
40	2.50	90	1.11	140	.71	190	.53
41	2.44	91	1.10	141	.71	191	.52
42	2.38	92	1.09	142	.70	192	.52
43	2.33	93	1.08	143	.70	193	.52
44	2.27	94	1.06	144	.69	194	.52
45	2.22	95	1.05	145	.69	195	.51
46	2.17	96	1.04	146	.68	196	.51
47	2.13	97	1.03	147	.68	197	.51
48	2.08	98	1.02	148	.68	198	.51
49	2.04	99	1.01	149	.67	199	.50
50	2.00	100	1.00	150	.67	200	.50

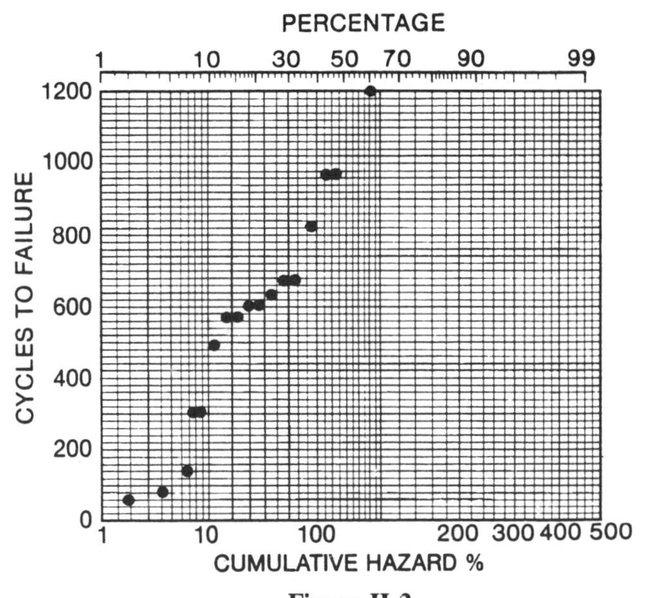

Figure II.2
Normal Hazard Plot of Appliance Component Data

How to Interpret a Hazard Plot

The probability and data scales on a hazard paper are exactly the same as those on the corresponding probability paper. Thus, a hazard plot is interpreted the same way as a probability plot, and the scales on hazard paper are used like those on probability paper. The cumulative hazard scale is only an aid for plotting multiply censored data.

Estimate of the percentage failing. The population percentage failing by a given age is estimated from the fitted line or curve as follows. Enter the plot on the time scale at the given age, go to the fitted line, and then go to the corresponding point on the probability scale to read the percentage. For example, the estimate of the percentage of components failing by 500 cycles (warranty) is 12%; this answers a basic question.

Percentile estimate. To estimate a percentile, enter the plot on the probability scale at the given percentage, go to the fitted line, and then go to the corresponding point on the time scale to read the percentile. For example, the estimate of the 50th percentile, nominal component life, is 1000 cycles.

3. LIFE DISTRIBUTION WITH FAILURE MODES ELIMINATED

Hazard plotting also provides an estimate of the life distribution that would result if certain failure modes were eliminated by proposed design changes. It is costly and time consuming to change a design and collect and analyze data to determine the value of design changes. Instead this can be done using past data. It is assumed that the cause of each failure is identified. The following example illustrates the method.

The method. Suppose that a proposed design change of the appliance would eliminate mode 11 failures and leave other failure modes unchanged. Past data including mode 11 and other failure modes are given in Table II.3 and are used to predict the resulting life distribution. Hazard calculations for the life distribution without mode 11 are shown on Table II.3. Each failure time by mode 11 is treated as a censoring time, since the new design would have run that long without failure. The failure times for the remaining modes are plotted against their cumulative hazard values as shown in Figure II.3 on Weibull paper. About 10% would fail on warranty (500 cycles) with mode 11 eliminated.

Failure rate behavior. Often it is useful to know how the failure rate depends on product age. A failure rate that increases with age usually indicates that failures are due to wearout. A failure rate that decreases with age usually indicates that failures are due to manufacturing or design defects that cause early failures.

For data plotted on Weibull hazard paper, the following assesses the nature of the failure rate. A Weibull failure rate increases (decreases) if the shape parameter is greater (less) than 1. To estimate the Weibull shape parameter, draw a straight line parallel to the plotted data, so it passes through the "origin" of the Weibull hazard paper and through the shape parameter scale, as in Figure II.3. The value on that scale is the estimate of the shape parameter; it is 0.7 in Figure II.3, indicating a decreasing failure rate (design or manufacturing defects).

Assumptions. The hazard plotting method above is based on four assumptions: 1) Each unit has a potential failure time for each failure mode. 2) The observed time to failure for a unit is the smallest of its potential times to failure. 3) Potential times to failure for different failure modes are statistically independent. 4) The mode of each failure is identified. Thus the product is regarded to be a series system.

4. LIFE DISTRIBUTION OF A FAILURE MODE

Information is sometimes desired on the distribution of time to failure for a particular failure mode. An estimate of its distribution provides information on the nature of the failure mode and on the effect of design changes on that mode.

The method. An example of the method involves the data with a mix of failure modes in Table II.3. Hazard calculations for mode 11 are shown in Table II.1. In these calculations, each failure time for another mode is treated as a censoring time for mode 11, that is, as if those units were removed from test before they failed by mode 11. The failure times for mode 11 are plotted against their cumulative hazard values in Figure II.2.

Table II.3
Hazard Calculations with Mode 11 Eliminated

Cycles	Failure Mode	Cum. Hazard	Cycles	Failure Mode	Cum. Hazard
45	1	1.85	608+		
47	11		608+		
73	11		608+		
136+			608+		
136	6	3.85	608+		
136+			608	11	
136+			608+		
136+			608+		
145	11		630	11	
190+			670	11	
190+			670	11	
281	12	6.18	731+		
311	11		838	11	
417	12	8.62	964	11	
485+			964	11	
485+			1164+		
490	11		1164+		
569	1	11.32	1164+		
571+			1164+		
571	11		1164+		
575	11		1164+		
608+			1164+		
608+			1198	9	31.32
608+			1198	11	
608+			1300+		
608+	11		1300+		
608+			1300+		

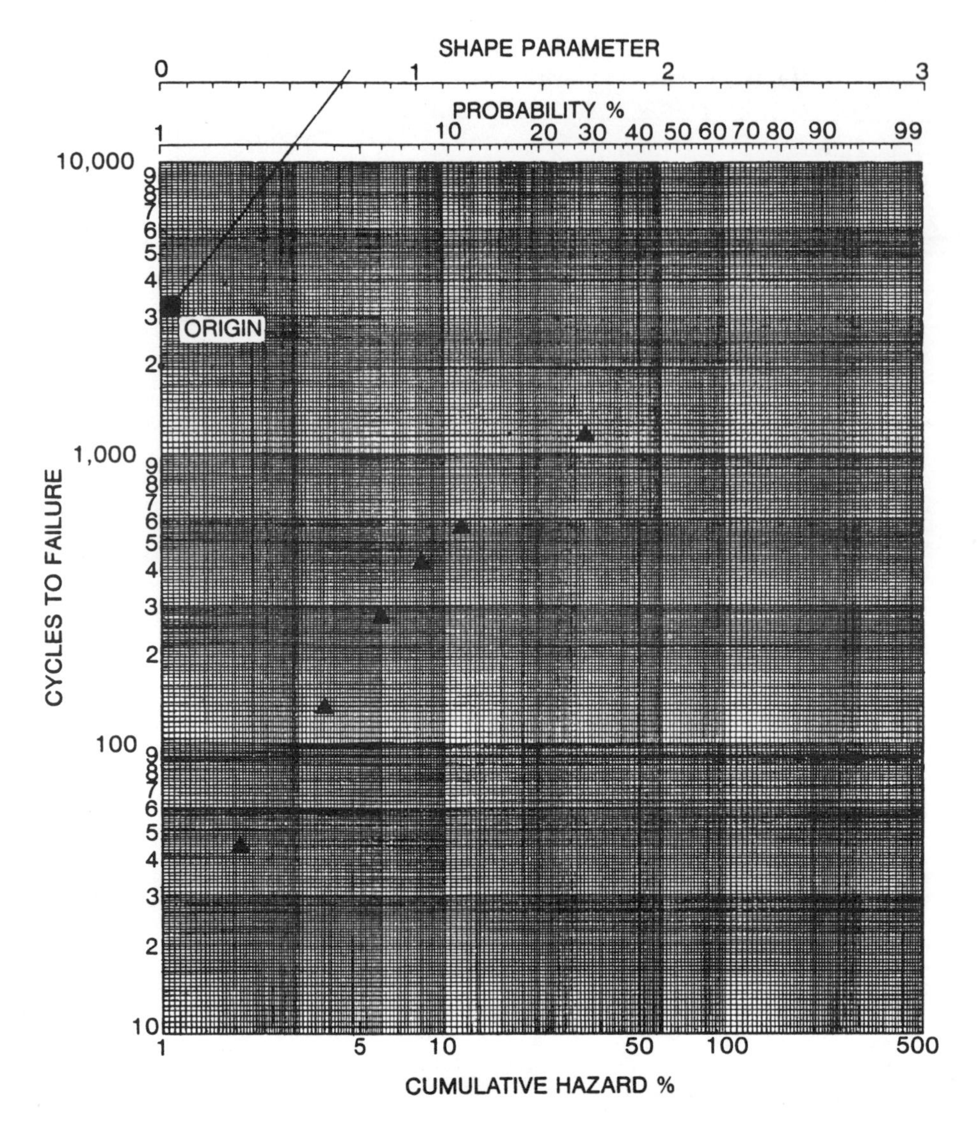

Figure II.3
Appliance Life with Mode 11 Eliminated

CHAPTER III NUMERICAL ANALYSES OF LIFE DATA

This chapter presents standard methods for analysis of Poisson (Sec. 1) and binomial (Sec. 2) data and maximum likelihood (ML) methods for censored and complete data from exponential (Sec. 3), normal and lognormal (Sec. 4), and Weibull distributions (Sec. 5), for data with competing failure modes (Sec. 6), and for inspection data (Sec. 7). Chapter IV contains a brief survey of other topics, including a list of books that provide further analyses for life data. Necessary background is Chapter I on basic concepts and distributions. Chapter II on graphical analyses is helpful.

While requiring sophisticated computer programs, ML methods are very important in life data analysis, because they apply to most theoretical distributions and types of data, particularly censored data. Also, ML estimators usually have good statistical properties. For example, for samples with many failures (say, over 20), the cumulative distribution function of most ML estimators is close to a normal one whose mean equals the true quantity being estimated and whose variance is as small as that of any other estimator. This normal distribution yields approximate confidence limits for the true value being estimated. An ML estimator also usually has good properties for small samples. Before using numerical methods, it is important to use a hazard plot to check the distribution and data. A combination of graphical and numerical methods often is most informative.

1. POISSON DATA

This section gives estimates and confidence intervals for the Poisson occurrence rate λ and guidance on sample size. Also, it gives a prediction and prediction limits for the number of occurrences in a future sample. Poisson data consist of the number Y of occurrences in a 'length' t of observation.

Estimate of λ

The estimate for the true λ is the sample occurrence rate

$$\hat{\lambda} = Y/t.$$

Its mean is λ, and

$$\mathrm{Var}(\hat{\lambda}) = \lambda/t.$$

Power line example. A new tree wire had $Y = 12$ failures in $t = 467.9$ thousand-feet-years of exposure. The estimate of the failure rate is $\hat{\lambda} = 12/467.9 = 0.0256$ failures per thousand-feet-years.

Confidence Interval for λ

Two-sided $100\gamma\%$ confidence limits for the true λ are

$$\underset{\sim}{\lambda} = 0.5\chi^2[(1-\gamma)/2; 2Y]/t, \qquad \tilde{\lambda} = 0.5\chi^2[(1+\gamma)/2; 2Y+2]/t;$$

here $\chi^2[\delta;\nu]$ is the 100δ-th chi-square percentile with ν degrees of freedom, tabled on pages 55 and 56. Often used in reliability work, the one-sided upper limit is

$$\tilde{\lambda} = 0.5\chi^2[\gamma; 2Y+2]/t.$$

For tree wire, two-sided 95% confidence limits are $\underset{\sim}{\lambda} = 0.5\cdot\chi^2\,[(1-0.95)/2;\ 2\cdot 12]/467.9 = 0.0133$ and $\tilde{\lambda} = 0.5\cdot\chi^2[(1+0.95)/2;\ 2\cdot 12+2]/467.9 = 0.0444$ failures per thousand-feet-years. Each limit is a one-sided 97.5% confidence limit.

Choice of t for Estimating λ

To choose the "sample size" t, the following is helpful. The estimate $\hat{\lambda}$ is within $\pm w$ of λ with approximate probability $100\gamma\%$ if

$$t=\lambda[K_\gamma/w]^2$$

where K_γ is the $100(1+\gamma)/2$-th standard normal percentile. Here one must approximate the unknown λ.

Suppose λ of tree wire is to be estimated within ± 0.0050 failures per thousand-feet-years with 95% probability. One needs approximately $t=0.0256(1.960/0.0050)^2=3{,}940$ thousand-feet-years.

Prediction

Often one seeks information on the random number of occurrences in a future sample. For example, those who maintain power lines need to predict the number of line failures in order to plan the number of repair crews.

The following gives a prediction and prediction limits for the number X of occurrences in a future observation of length s with true rate λ. Suppose past data consist of Y occurrences in an observation of length t with the same true rate λ.

The prediction of the future number is the observed rate $\hat{\lambda}=Y/t$ times the future length s of observation:

$$\hat{X}=\hat{\lambda}s=(Y/t)s.$$

If all power lines in the region were tree wire, there would be $s=515.8$ thousand feet. Then the prediction of the number of failures in the coming year would be $\hat{X}=(12/467.9)515.8\cong 13.2$ failures, helpful for maintenance planning.

Prediction Limits

Two-sided approximate $100\gamma\%$ prediction limits for X are

$$\underset{\sim}{X}\cong\hat{X}-K_\gamma[\hat{\lambda}s(t+s)/t]^{1/2},\quad \tilde{X}\cong\hat{X}+K_\gamma[\hat{\lambda}s(t+s)/t]^{1/2};$$

here K_γ is the $100(1+\gamma)/2$-th standard normal percentile. Y and $\hat{\lambda}s$ should both exceed 10 for an adequate approximation. For such a one-sided limit, replace K_γ by z_γ, the 100γ-th standard normal percentile. Nelson (1970) gives simple charts for exact limits.

Two-sided approximate 95% prediction limits for the number of tree wire failures next year are $\underset{\sim}{X}=13.2-1.960[(0.0256)515.8(467.9+515.8)/467.9]^{1/2}\cong 3$ failures and $\tilde{X}=13.2+1.960[(0.0256)515.8(467.9+515.8)/467.9]^{1/2}\cong 24$ failures. Here $Y=12>10$ and $\hat{\lambda}s=0.0256(515.8)=13.2>10$; so the approximation is adequate.

2. BINOMIAL DATA

This section presents estimates and confidence intervals for the binomial probability p and guidance on sample size. Also, it presents a prediction and prediction limits for the number of failures in a future sample. Binomial data consist of the number Y of "category" units among n statistically independent sample units, each with the same probability p of being a category unit.

Estimate of p

The estimate of the population proportion p is the sample proportion

$$\hat{p}=Y/n.$$

Its mean equals p, and

$$\text{Var}\,(\hat{p})=p(1-p)/n.$$

If np and $n(1-p)$ exceed 10, the distribution of $\hat{p}$ is approximately normal with a mean of p and variance above.

Locomotive control example. Of $n=96$ locomotive controls on test, $Y=15$ failed on warranty. The estimate of the population proportion failing on warranty is $\hat{p}=15/96=0.156$ or 15.6%.

Confidence Interval for p

Two-sided $100\gamma\%$ confidence limits for p are

$$\underset{\sim}{p}=\{1+(n-Y+1)Y^{-1}F[(1+\gamma)/2;2n-2Y+2,2Y]\}^{-1},$$
$$\tilde{p}=\{1+(n-Y)\{(Y+1)F[(1+\gamma)/2;2Y+2,2n-2Y]\}^{-1}\}^{-1};$$

here $F[\delta;a,b]$ is the 100δ-th F-percentile with a degrees of freedom in the numerator and b in the denominator. Approximate limits are

$$\underset{\sim}{p}\cong\hat{p}-K_\gamma[\hat{p}(1-\hat{p})/n]^{1/2},\quad \tilde{p}\cong\hat{p}+K_\gamma[\hat{p}(1-\hat{p})/n]^{1/2}.$$

Here K_γ is the $100(1+\gamma)/2$-th standard normal percentile. The approximation is satisfactory if $Y>10$ and $n-Y>10$.

Obtaining exact limits is easier with the Clopper-Pearson charts in Figures III.1A and B. Enter a chart on the horizontal axis at $\hat{p}=Y/n$; go vertically to the curve labeled n (separate curves for the lower and upper limits); and go horizontally to the vertical scale to read the confidence limit.

For a 95% confidence interval for the locomotive control, enter the 95% chart on the horizontal axis at $\hat{p}=15/96=0.156$; go vertically to a curve for $n=96$ (interpolate); and go horizontally to the vertical scale to read $\underset{\sim}{p}=9\%$ and $\tilde{p}=24\%$. Each limit is a one-sided 97.5% confidence limit for p.

Sample Size n to Estimate p

The estimate $\hat{p}$ is within $\pm w$ of p with approximately $100\gamma\%$ probability if

$$n\cong p(1-p)(K_\gamma/w)^2;$$

here K_γ is the $100(1+\gamma)/2$-th standard normal percentile. To use this, one must approximate the unknown p. Otherwise, the largest sample size ($p=1/2$) is

$$n\cong 0.25(K_\gamma/w)^2.$$

These formulas are usually accurate enough if $np>10$ and $n(1-p)>10$.

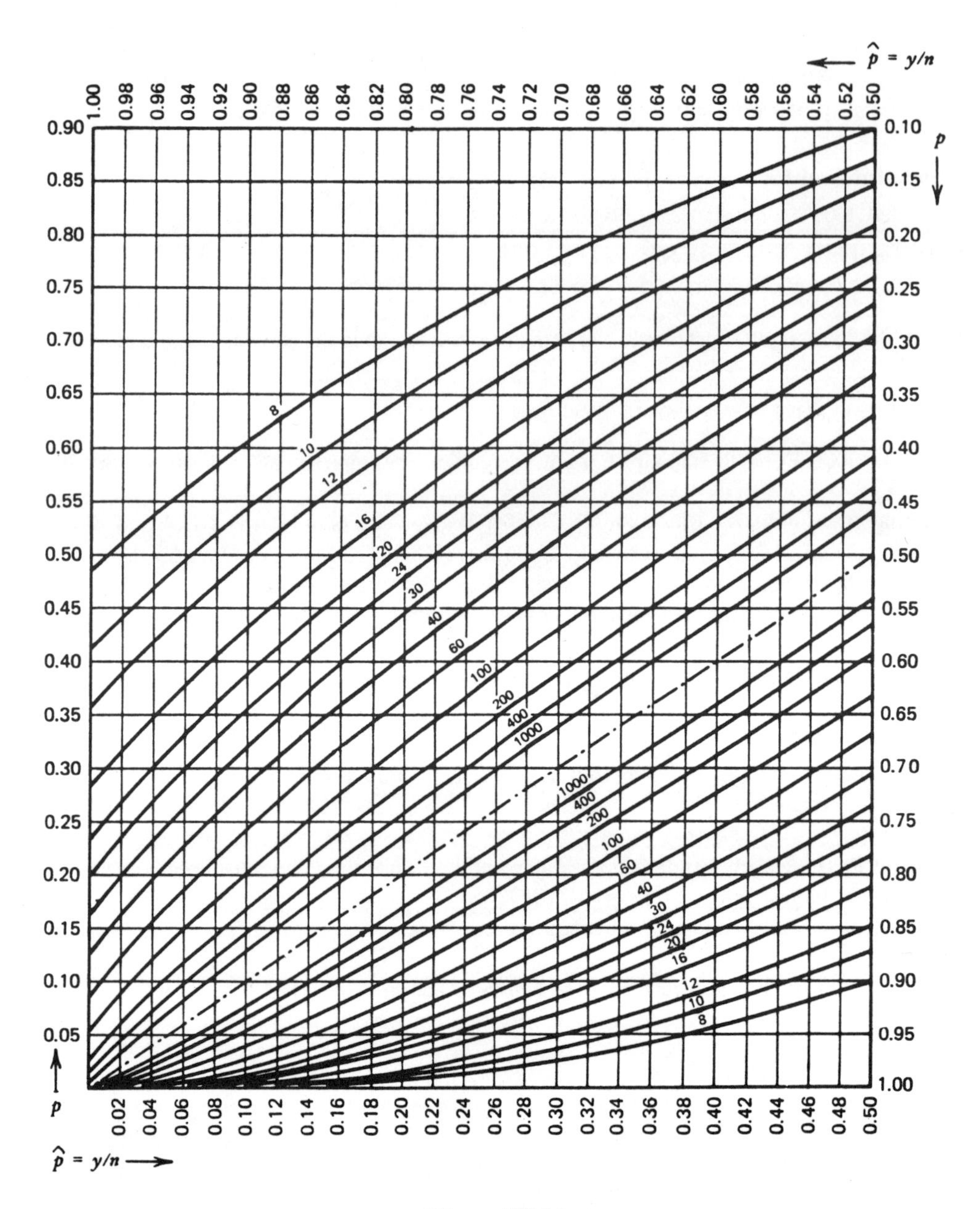

Figure III.1A
Confidence Limits for Binomial p – 99% Two-Sided or 99.5% One-Sided

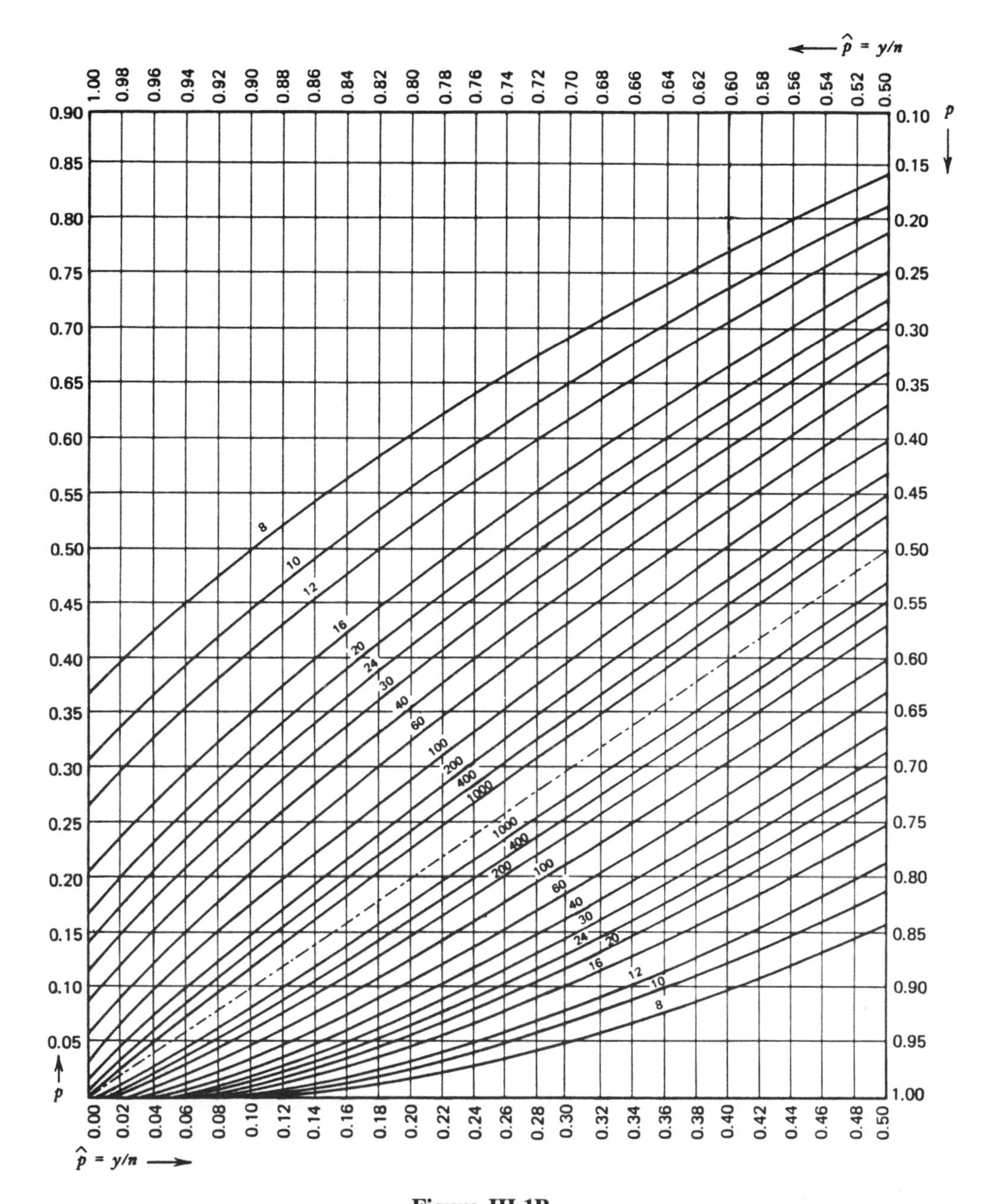

Figure III.1B
Confidence Limits for Binomial p — 95% Two-Sided or 97.5% One-Sided.
From G.E.P. Box, W.G. Hunter, and J.S. Hunter, *Statistics for Experimenters*, Wiley, New York, 1978, pp. 642-643. Reproduced by permission of the publisher and the Biometrika Trustees.

Suppose the percentage of locomotive controls failing on warranty is to be estimated within $\pm 5\%$ (w=0.05) with 95% probability. This requires $n \cong 0.156(1-0.156)(1.960/0.05)^2=202$ controls; $\hat{p}$ was used for the unknown p. The largest $n \cong 0.25(1.960/0.05)^2=384$ controls. Here $n\hat{p}=202(0.156)=31.5>10$, and $n(1-\hat{p})=202(1-0.156)=170.5>10$. So the approximation is satisfactory.

Prediction

The following provides a prediction and prediction limits for the number X of category units in a future sample of m units. Suppose the past sample consists of Y category units in a sample of n units from the same population.

The prediction of the number X of future category units is

$$\hat{X}=m\hat{p}=m\ (Y/n).$$

The prediction error $\hat{X}-X$ has a mean of zero, and the prediction error variance is

$$\text{Var}\ (\hat{X}-X)=mp\ (1-p)\ (m+n)/n;$$

here p is the unknown true population proportion.

Locomotive control example. Suppose future production is m=900 controls. For Y=15 failures among n=96 controls, the prediction of the number of future warranty failures is $\hat{X}=900(15/96)=141$ controls.

Prediction Limits

Lieberman and Owen (1961) give exact prediction limits for X, which are laborious. Two-sided approximate $100\gamma\%$ prediction limits for X are

$$\underset{\sim}{X} \cong \hat{X}-K_\gamma\ [m\hat{p}\ (1-\hat{p})\ (m+n)/n]^{1/2}, \quad \tilde{X} \cong \hat{X}+K_\gamma[m\hat{p}\ (1-\hat{p})\ (m+n)/n]^{1/2};$$

here K_γ is the 100 $(1+\gamma)/2$-th standard normal percentile. These limits are usually accurate enough if Y, $n-Y$, $\hat{X}$, and $m-\hat{X}$ all exceed 10.

For the locomotive control, two-sided 95% prediction limits are

$$\underset{\sim}{X} \cong 141-1.960[(900)0.156(1-0.156)(900+96)/96]^{1/2}=72,$$

$$\tilde{X} \cong 141+1.960[(900)0.156(1-0.156)(900+96)/96]^{1/2}=210.$$

Each limit is a one-sided 97.5% prediction limit. $Y=15$, $n-Y=81$, $\hat{X}=141$, and $m-\hat{X}=759$ exceed 10; so the limits are accurate enough.

3. EXPONENTIAL DATA

This section presents maximum likelihood (ML) methods for fitting an exponential distribution to multiply censored data. Other methods for exponential data appear in the books listed in Section 12 of Chapter IV. Suppose the n sample times are $y_1, y_2, \ldots, y_n$ of which r are failure times and n-r are survival times.

ML Estimate for the Mean θ

The ML estimate $\hat{\theta}$ for the true population mean θ is

$$\hat{\theta} = \sum_{i=1}^{n} y_i/r.$$

This is the total time on all n units divided by the number r of failures.

Engine fan example. Table III.1 shows data on n=70 diesel engine fans that accumulated 344,440 hours in service while r=12 failed. Management wanted an estimate and confidence limits (given later) for the fraction failing on an 8000 hour warranty. This information was needed to determine whether to redesign the fan. The ML estimate of θ is $\hat{\theta}$=344,440/12=28,700 hours. Figure III.2 is a Weibull hazard plot of the data. The ML fit of the exponential distribution is the center line on the plot; the slope of the center line corresponds to a Weibull shape parameter of β=1. The other two lines are 95% confidence limits for exponential percentiles (and reliabilities), described later.

Confidence Limits for θ

Approximate 100γ% *confidence limits* for θ are

$$\underset{\sim}{\theta} \cong \hat{\theta}\ 2r/\chi^2[(1+\gamma)/2;2r],\quad \tilde{\theta} \cong \hat{\theta}\ 2r/\chi^2[(1-\gamma)/2;2r];$$

here $\chi^2(\delta;2r)$ is the 100δ-th chi-square percentile with $2r$ degrees of freedom, tabled on pages 55 and 56. For a one-sided 100γ% confidence limit, replace $(1+\gamma)/2$ by γ or $(1-\gamma)/2$ by $1-\gamma$.

Engine fan. Two-sided approximate 95% confidence limits for the fan mean life are

$$\underset{\sim}{\theta} \cong 28{,}700 \times 2 \times 12/\chi^2[0.975;2\times 12] = 28{,}700\{24/39.36\} = 17{,}500 \text{ hours},$$

$$\tilde{\theta} \cong 28{,}700 \times 2 \times 12/\chi^2[0.025;2\times 12] = 28{,}700\{24/12.40\} = 55{,}500 \text{ hours}.$$

Samples with no failures. For samples with no failures, a conservative commonly used one-sided lower 100γ% confidence limit for θ uses r=1 and is

$$\underset{\sim}{\theta} = 2(y_1 + \ldots + y_n)/\chi^2(\gamma;2) = -(y_1 + \ldots + y_n)\ln(1-\gamma).$$

For an estimate, some use a 50% confidence limit; this estimate is conservatively biased toward low values and can be seriously pessimistic.

Sample Size

One may wish to choose a sample size to achieve a confidence interval of a desired length. One measure of length is the ratio $\tilde{\theta}/\underset{\sim}{\theta} = \chi^2[(1+\gamma)/2;2r]/\chi^2[(1-\gamma)/2;2r]$; this is a function only of the number of failures r, not the sample size n. The ratio can be calculated for a number of r values and a suitable r chosen. Of course, the larger n is, the more quickly the r failures occur. In particular, the expected test time to the r-th failure in a singly censored sample of size n is $\theta[n^{-1}+(n-1)^{-1}+\ldots+(n-r+1)^{-1}] \cong \theta \cdot \ln[(n+½)/(n-r+½)]$.

Exponential Failure Rate λ

The ML estimate of the true failure rate $\lambda=1/\theta$ is

$$\tilde{\lambda} = 1/\tilde{\theta} = r \Big/ \sum_{i=1}^{n} y_i.$$

This is the number r of failures divided by the total accumulated time, the "sample failure rate." One- or two-sided 100γ% confidence limits for λ are

$$\underset{\sim}{\lambda} = 1/\tilde{\theta},\quad \tilde{\lambda} = 1/\underset{\sim}{\theta};$$

here $\underset{\sim}{\theta}$ and $\tilde{\theta}$ are the one- or two-sided limits above.

Table III.1
Fan Failure Data (Hours)

450	1850+	2200+	3750+	4300+	6100+	7800+	8750
460+	1850+	3000+	4150+	4600	6100	7800+	8750+
1150	1850+	3000+	4150+	4850+	6100+	8100+	9400+
1150	2030+	3000+	4150+	4850+	6100+	8100+	9900+
1560+	2030+	3000+	4150+	4850+	6300+	8200+	10100+
1600	2030+	3100	4300+	4850+	6450+	8500+	10100+
1660+	2070	3200+	4300+	5000+	6450+	8500+	10100+
1850+	2070	3450	4300+	5000+	6700+	8500+	11500+
1850+	2080	3750+		5000+	7450+	8750+	

+ Denotes running time.

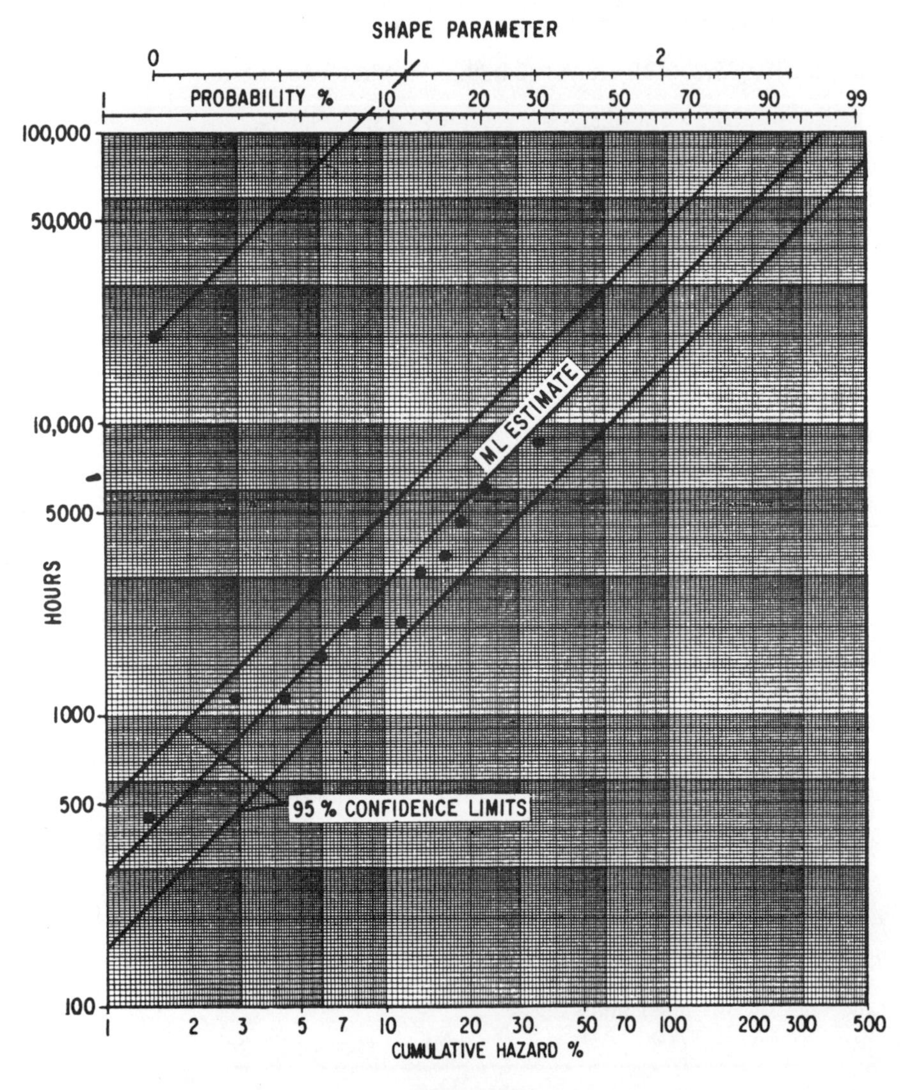

Figure III.2
Weibull Hazard Plot of Fan Failures — Exponential Fit

Engine fan. For the fans, $\hat{\lambda}=(12/344{,}440)/10^6=34.8$ failures per million hours. For 95% confidence, $\underset{\sim}{\lambda}\cong 10^6/55{,}500=18.0$ and $\tilde{\lambda}\cong 10^6/17{,}500=57.1$ failures per million hours. Each limit is a one-sided 97.5% confidence limit.

Exponential Percentile

The 100P-th exponential percentile is $y_P=\ -\theta\ln(1-\mathrm{P})$. Its ML estimate is

$$\hat{y}_P=\ -\hat{\theta}\ln(1-P).$$

One- or two-sided 100γ% confidence limits for y_P are

$$\underset{\sim}{y}_P=-\underset{\sim}{\theta}\ln(1-P),\quad \tilde{y}_P=-\tilde{\theta}\ln(1-P).$$

Engine fan. The ML estimate of the 10th percentile of fan life is $\hat{y}_{.10}=-28{,}700\cdot\ln(1-0.10)=3020$ hours. For 95% confidence, $\underset{\sim}{y}_{.10}=-17{,}500\cdot\ln(1-0.10)=1840$ and $\tilde{y}_{.10}=-55{,}500\cdot\ln(1-0.10)=5850$ hours. The ML line in Figure III.2 gives the ML estimates of the exponential percentiles. On Weibull paper (Figure III.2), confidence limits for exponential percentiles are straight lines parallel to the ML line.

Exponential Reliability

Reliability at age y is $R(y)=\exp(-y/\theta)$. Its ML estimate is

$$\hat{R}(y)=\exp(-y/\hat{\theta}).$$

One- or two-sided 100γ% confidence limits for $R\,(y)$ are

$$\underset{\sim}{R}(y)=\exp(-y/\underset{\sim}{\theta}),\quad \tilde{R}(y)=\exp(-y/\tilde{\theta}).$$

The corresponding estimate and limits for the fraction failing, $F(y)=1-R(y)$, are

$$\hat{F}(y)=1-\hat{R}(y),\quad \underset{\sim}{F}(y)=1-\tilde{R}(y),\quad \tilde{F}(y)=1-\underset{\sim}{R}(y).$$

Engine fan. The ML estimate of fan reliability on an 8000 hour warranty is $\hat{R}(8000)=\exp(-8000/28{,}700)$ $=0.76$. Approximate 95% confidence limits are $\underset{\sim}{R}(8000 = \exp(-8000/17{,}500)=0.63$ and $\tilde{R}(8000)=\exp$ $(-8000/55{,}500)=0.87$. For the fraction failing, $\hat{F}(8000)=0.24$, $\underset{\sim}{F}(8000)=0.13$, and $\tilde{F}(8000)=0.37$. Management used this information to decide whether to replace unfailed fans with an improved fan. In Figure III.2, the straight-line confidence limits for percentiles are also the confidence limits for failure probabilities.

Computer Programs

Computer programs that ML fit an exponential distribution to data are not necessary, as hand calculations are easy. Such programs include STATPAC by Nelson and others (1978) and SURVREG by Preston and Clarkson (1980). Each gives the ML estimates and confidence limits for the exponential mean, percentiles, and failure probabilities (including reliabilities).

4. WEIBULL DATA

This section presents maximum likelihood (ML) fitting of a Weibull distribution to multiply time censored data. Standard computer programs yield ML estimates and approximate confidence limits for parameters, percentiles, and reliability. Other methods for Weibull data appear in the books listed in Section 12 of Chapter IV.

Maximum Likelihood Estimates

For a sample of n units, denote the r failure times by $t_1, t_2, \ldots, t_r$ and the n-r survival times by $t_{r+1}, \ldots, t_n$. Then the ML estimate $\hat{\beta}$ of the shape parameter β is the solution of

$$\sum_{i=1}^{r} \ln(t_i)/r = \left[\sum_{i=1}^{n} t_i^{\beta}\ln(t_i)\right]\left[\sum_{i=1}^{n} t_i^{\beta}\right]^{-1} - (1/\beta).$$

Computer programs iteratively solve this equation to get $\hat{\beta}$. The ML estimate $\hat{\alpha}$ for the characteristic life α is

$$\hat{\alpha} = \left[\sum_{i=1}^{n} t_i^{\hat{\beta}}/r\right]^{1/\hat{\beta}}$$

Calculations of approximate confidence limits for β, α, percentiles, and reliabilities are too complex to give here but appear in Nelson (1982). Such confidence limits are calculated by some Weibull programs.

Engine Fan Example

Section 3 presents an exponential fit to the data on 70 engine fans in Table III.1. The following presents a more general Weibull fit. Management wanted an estimate and confidence limits for the fraction failing on an 8000 hour warranty. Also, management wanted to know if fans had an increasing or decreasing failure rate; this would indicate whether to replace first new or old unfailed fans with a new durable design to avoid service failures.

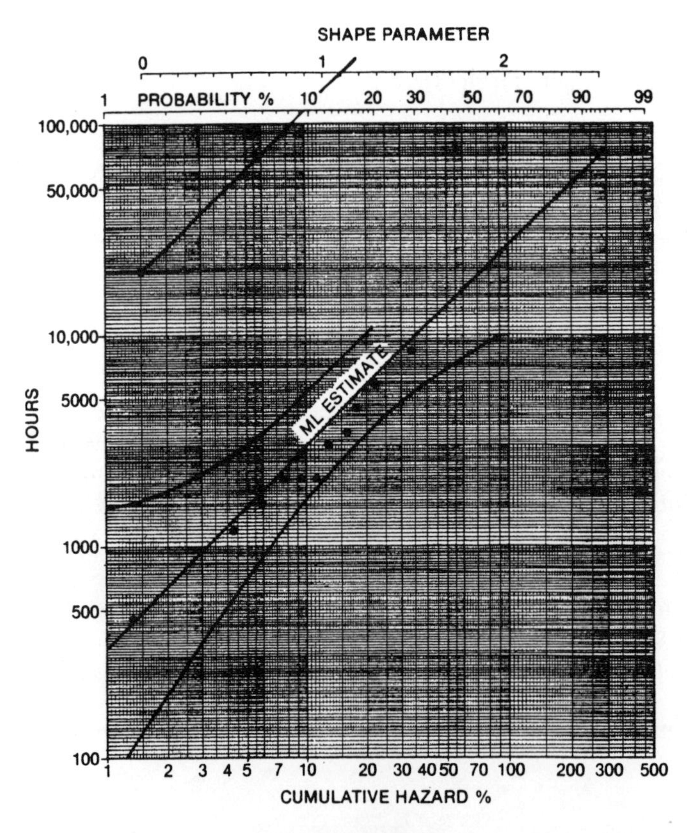

Figure III.3
Weibull Hazard Plot of Fan Failures — Weibull Fit and 95% Confidence Limits

Figure III.3 is a Weibull hazard plot of the data. The ML fit of the Weibull distribution is the straight line in the plot, and the curves are approximate 95% confidence limits for percentiles. In Figure III.2 for the exponential fit, the confidence limits are closer together and are straight lines.

Figure III.4 shows ML output of the WEIB program of the General Electric Information Service Co. (1979). ML estimates are $\hat{\alpha}=26.3$ thousand hours and $\hat{\beta}=1.06$, which is close to 1 and indicates that the failure rate is essentially constant. That is, new and old fans are about equally failure prone. The approximate 95% confidence limits $\underset{\sim}{\beta}=0.64$ and $\tilde{\beta}=1.74$ enclose 1. Thus the data are consistent with $\beta=1$, corresponding to the exponential distribution and a constant failure rate. So management had no convincing indication to replace old fans before new ones. Replacements could be made in the most convenient and economical way.

From Figure III.4, the estimate of the fraction failing on an 8 thousand-hour warranty is $\hat{F}(8)=1-0.75=0.25$. Approximate 95% confidence limits are $\underset{\sim}{F}(8)=1-0.86=0.14$ and $\tilde{F}(8)=1-0.61=0.39$. This helped management decide to replace unfailed fans with a more durable fan.

Tables for exact confidence limits for Weibull parameters, percentiles, and reliabilities for *singly* censored data are given by McCool (1974), Bain (1978), and Lawless (1978). Exact limits are recommended for samples with few failures (say, 10 or fewer).

Computer Programs

Computer programs that ML fit the Weibull distribution to multiply censored data include:

- STATPAC by Nelson and others (1978) gives the ML estimates and approximate confidence limits for parameters (α,β), percentiles, and probabilities (including reliabilities). It runs on Honeywell 600/6000's.
- WEIB of the General Electric Information Service Co. (1979) gives the same results as STATPAC. Figure III.4 shows output from WEIB.
- CENSOR by Meeker and Duke (1979) is similar to STATPAC, costs less, but is less easy to use. In Fortran IV, it runs on IBM and similar computers.
- SURVREG by Preston and Clarkson (1980) is similar to STATPAC. In Fortran IV, it runs on AMDAHL, IBM, and similar computers.
- Many organizations have simple computer programs that merely calculate the ML estimates of the Weibull parameters but not confidence intervals.

5. NORMAL AND LOGNORMAL DATA

This section presents maximum likelihood (ML) fitting of normal and lognormal distributions to multiply censored data. Formulas for ML estimates and confidence limits for μ and σ are complex and appear in Nelson (1982). Computer programs provide ML estimates and approximate confidence limits for parameters μ and σ, percentiles, and reliabilities. Other methods for normal and lognormal data appear in the books listed in Section 12 of Chapter IV.

Locomotive Control Example

Table III.2 shows singly time censored life data on 96 locomotive controls from Hahn and Shapiro (1967). Management wanted an estimate and confidence limits for the fraction of controls failing on an 80 thousand-mile warranty. Figure III.5 is a lognormal probability plot of the data. The ML fit of the lognormal distribution is the straight line on the plot; the curved lines are approximate 95% confidence limits for the distribution percentiles (and reliabilities).

```
      ESTIMATE AND TWO-SIDED 95.00% CONFIDENCE
       INTERVALS FOR DISTRIBUTION PARAMETERS

  SHAPE (BETA) PARAMETER:                     1.0584
  LOWER LIMIT:                               0.64408
  UPPER LIMIT:                                1.7394

  SCALE (63.2 PCTILE) PARAMETER:              26.297
  LOWER LIMIT:                                10.552
  UPPER LIMIT:                                65.534


  ESTIMATED COVARIANCE MATRIX OF PARAMETER ESTIMATES:

                    SCALE              SHAPE
     SCALE         150.10            -2.6645
     SHAPE        -2.6645         0.71959E-01

          ESTIMATE AND TWO-SIDED 95.00% CONFIDENCE
          INTERVALS FOR DISTRIBUTION PERCENTILES
```

PERCEN-TAGE	PERCENTILE ESTIMATE	LOWER LIMIT	UPPER LIMIT
0.10	0.38527E-01	0.29850E-02	0.49726
0.50	0.17659	0.28467E-01	1.0954
1.00	0.34072	0.74824E-01	1.5515
5.00	1.5893	0.68311	3.6974
10.00	3.1372	1.6862	5.8369
20.00	6.3747	3.7305	10.893
25.00	8.1039	4.6441	14.141
50.00	18.600	8.5248	40.584
75.00	35.804	12.639	101.43
80.00	41.225	13.696	124.09
90.00	57.825	16.541	202.16
95.00	74.147	18.949	290.14
99.00	111.31	23.578	525.46
99.50	127.07	25.301	638.22
99.90	163.27	28.891	922.64

```
WANT AN ESTIMATE OF PROBABILITY OF SURVIVAL BEYOND A SPECIFIC TIME?
IF SO, TYPE A DESIRED TIME; OTHERWISE, TYPE 0 --? 8


          ESTIMATE AND TWO-SIDED 95.00% CONFIDENCE
    INTERVAL FOR THE DISTRIBUTION PERCENTAGE ABOVE        8.000

          ESTIMATE:                             0.75293
          LOWER LIMIT:                          0.60817
          UPPER LIMIT:                          0.85681
```

Figure III.4
WEIB Output for Fan Failure Data

Maximum Likelihood Fit

Figure III.6 shows computer output from STATPAC of Nelson and others (1978). The output gives ML estimates and approximate confidences limits for the parameters, selected percentiles, and the fraction failing on warranty. The ML estimates of the lognormal parameters are $\hat{\mu}=2.2223$ and $\hat{\sigma}=0.3064$ (base 10 logs). The estimate of median (50%) life is 167 thousand-miles, a "typical" life.

Figure III.6 shows that the ML estimate of the percentage failing on warranty is $\hat{F}(80)=15\%$. Approximate 95% confidence limits are $\underset{\sim}{F}(80)=10\%$ and $\tilde{F}(80)=22\%$. Management decided that the control must be redesigned, as warranty costs would be too high.

Tables for exact confidence limits for (log) normal parameters from *singly* censored data are given by Schmee, Gladstein, and Nelson (1982). Exact limits are recommended for samples with 10 or fewer failures.

Table III.2
Thousands of Miles to Failure for Locomotive Controls

22.5	57.5	78.5	91.5	113.5	122.5
37.5	66.5	80.0	93.5	116.0	123.0
46.0	68.0	81.5	102.5	117.0	127.5
48.5	69.5	82.0	107.0	118.5	131.0
51.5	76.5	83.0	108.5	119.0	132.5
53.0	77.0	84.0	112.5	120.0	134.0
54.5					

59 controls ran 135,000 miles without failure.

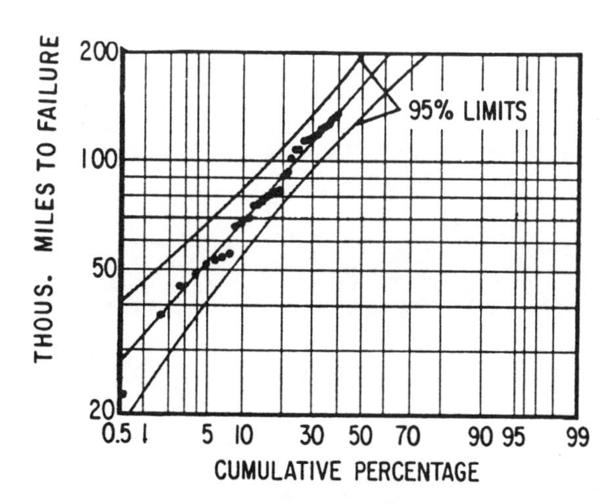

Figure III.5
Lognormal Probability Plot of Locomotive Control Failures

```
* MAXIMUM LIKELIHOOD ESTIMATES FOR DIST. PARAMETERS
  WITH APPROXIMATE 95% CONFIDENCE LIMITS

PARAMETERS      ESTIMATE       LOWER LIMIT       UPPER LIMIT

CENTER μ̂        2.222269      μ̰ = 2.133593      μ̃ = 2.310946
SPREAD σ̂       0.3064140      σ̰ =0.2365178      σ̃ =0.3969661

* COVARIANCE MATRIX

PARAMETERS      CENTER             SPREAD
                  μ̂                  σ̂
CENTER μ̂    0.2046921E-02
SPREAD σ̂    0.1080997E-02     0.1638384E-02

PCTILES

* MAXIMUM LIKELIHOOD ESTIMATES FOR  DIST.  PCTILES
  WITH APPROXIMATE 95% CONFIDENCE LIMITS

PCT.             ESTIMATE       LOWER LIMIT      UPPER LIMIT
100P               ŷP               y̰P               ỹP
0.1             18.84908         11.73800         30.26819
0.5             27.09410         18.40041         39.89532
1               32.30799         22.85237         45.67604
5               52.25811         40.94096         66.70360
10              67.53515         55.28316         82.50246
20              92.13728         77.87900         109.0060
50             >166.8282         136.0171         204.6188
80              302.0672         219.3552         415.9673
90              412.1062         278.7152         609.3370
95              532.5805         338.8504         837.0715
99              861.4479         487.2066         1523.157


PERCENT(LIMIT 80.) F(80)

* MAXIMUM LIKELIHOOD ESTIMATES FOR  % WITHIN LIMITS
  WITH APPROXIMATE 95% CONFIDENCE LIMITS

                  ESTIMATE          LOWER LIMIT         UPPER LIMIT

        PCTF̂(80)=14.87503     F̰(80) =9.895509     F̃(80) = 21.75530
```

Figure III.6
STATPAC Output on Lognormal Fit to Control Data

Computer Programs

Computer programs that ML fit the normal and lognormal distributions to multiply censored data include:

- STATPAC by Nelson and others (1978). It gives the ML estimates and approximate confidence limits for normal and lognormal parameters (μ,σ), percentiles, and probabilities (including reliabilities). It runs on Honeywell 600/6000's. Figure III.6 shows STATPAC output.
- CENSOR by Meeker and Duke (1979) is similar to STATPAC, costs less, but is less easy to use. It runs on IBM and similar computers.
- CENS by Hahn and Miller (1968) gives ML estimates and confidence limits for the parameters.
- Glasser's (1965) program is similar to that of Hahn and Miller (1968).
- IMSL (1975) OTMLNR program. It runs on many computers.
- SURVREG by Preston and Clarkson (1980). It runs on AMDAHL, IBM, and similar computers.

6. DATA WITH COMPETING FAILURE MODES

This section describes maximum likelihood (ML) methods for data where the mode (cause) of each failure is *known*. The methods estimate:

1) a separate life distribution for each failure mode,
2) the life distribution when all modes act, and
3) the life distribution that would result if certain modes are eliminated.

Chapter II presents hazard plotting of such data.

Connection Data

The data in Table III.3A are breaking strengths of 20 wire connections, one end bonded to a semiconductor wafer and the other to a terminal post. Each failure resulted from breaking of the wire or a bond (whichever is weaker). One problem was to estimate the strength distribution of connections, as a specification required that less than 1% have strength below 500 mg. Another problem was to estimate the strength distribution that would result from a redesign that eliminates 1) wire failures or else 2) bond failures. Strength takes the place of time to failure in this example.

Figure III.7A shows a normal hazard plot of all 20 strengths, ignoring cause of failure. Figure III.7B is a plot of the wire failures, and Figure III.7C is a plot of the bond failures.

The Model

The model used here for a product with competing failure modes is the series-system model of Chapter I.

Analysis of Data on a Single Failure Mode

Each sample unit has a time to failure by, say, mode 1 or else a time without mode 1. A unit that fails by another mode is treated as a censoring time for mode 1, since its failure time for mode 1 is beyond. Then the failure and censoring times for mode 1 are a multiply censored sample, and the distribution for mode 1 can be estimated with ML methods. For each mode, a separate distribution can be fitted to the failure and censoring times. Moreover, the ML estimates of the distribution parameters for different modes are statistically independent for large samples.

Such data on the wire strength appear in Table III.3B. Table III.3C shows such bond data.

The normal distribution was fitted by STATPAC to the data for each failure mode. Figures III.7A and B show the fitted distributions and 95% confidence limits. Figures III.8A and B show the STATPAC estimates and confidence intervals for parameters and percentiles.

Table III.3A
Connection Strength Data

Strength	Break of	Strength	Break of
550	B	1250	B
750	W	1350	W
950	B	1450	B
950	W	1450	B
1150	W	1450	W
1150	B	1550	B
1150	B	1550	W
1150	W	1550	W
1150	W	1850	W
1250	B	2050	B

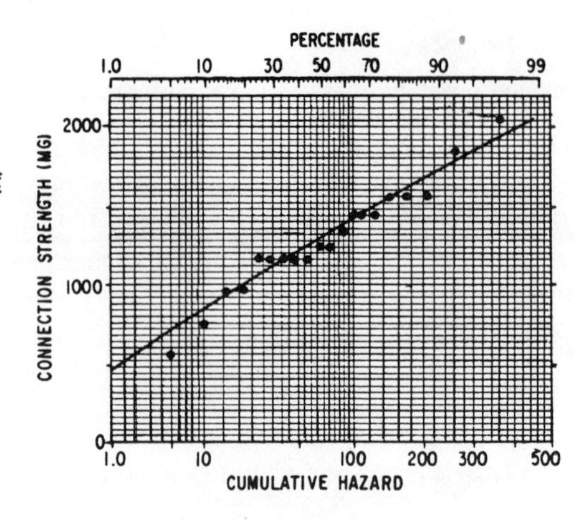

Figure III.7A
Hazard Plot of Connection Strengths

Table III.3B
Wire Strength Data

550+	1150+	1250+	1550+
750	1150+	1350	1550
950+	1150	1450+	1550
950	1150	1450+	1850
1150	1250+	1450	2050+

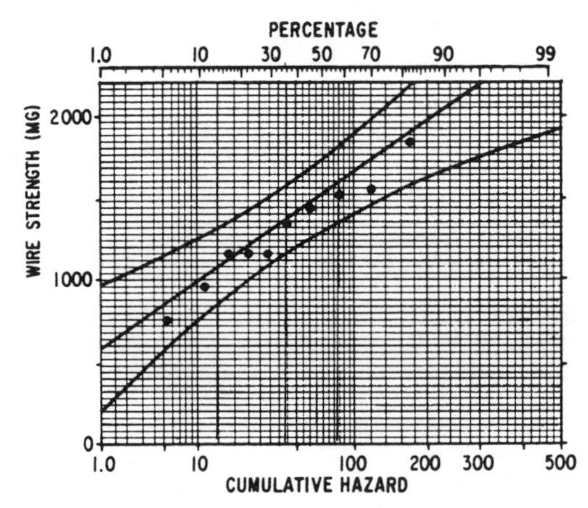

Figure III.7B
Hazard Plot of Wire Strengths

Table III.3C
Bond Strength Data

550	1150	1250	1550
750+	1150	1350+	1550+
950	1150+	1450	1550+
950+	1150+	1450	1850+
1150+	1250	1450+	2050

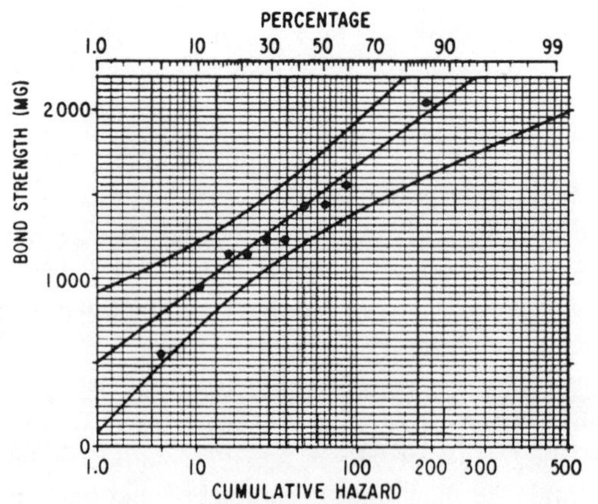

Figure III.7C
Hazard Plot of Bond Strengths

```
* MAXIMUM LIKELIHOOD ESTIMATES FOR DIST. PARAMETERS
  WITH APPROXIMATE 95% CONFIDENCE LIMITS
```

PARAMETERS	ESTIMATE	LOWER LIMIT	UPPER LIMIT
CENTER	$\hat{\mu}_W$ = 1517.384	$\underset{\sim}{\mu}_W$ = 1298.909	$\tilde{\mu}_W$ = 1735.859
SPREAD	$\hat{\sigma}_W$ = 398.8265	$\underset{\sim}{\sigma}_W$ = 256.3974	$\tilde{\sigma}_W$ = 620.3751

```
* COVARIANCE MATRIX
```

PARAMETERS	CENTER $\hat{\mu}_W$	SPREAD $\hat{\sigma}_W$
CENTER $\hat{\mu}_W$	12424.89	
SPREAD $\hat{\sigma}_W$	3606.373	8081.704

```
PCTILES

* MAXIMUM LIKELIHOOD ESTIMATES FOR DIST. PCTILES
  WITH APPROXIMATE 95% CONFIDENCE LIMITS
```

PCT. P	ESTIMATE $\hat{y}_P$	LOWER LIMIT $\underset{\sim}{y}_P$	UPPER LIMIT $\tilde{y}_P$
0.1	284.8018	-223.7595	793.3631
0.5	489.9127	62.82445	917.0010
1	589.4003	200.3698	978.4307
5	861.2300	567.6657	1154.794
10	1006.197	754.7586	1257.634
20	1181.789	966.3855	1397.192
50	1517.384	1298.909	1735.859
80	1852.979	1547.971	2157.988
90	2028.571	1662.169	2394.973
95	2173.538	1752.405	2594.670
99	2445.367	1915.946	2974.789

```
PERCENT(500. LIMIT)

* MAXIMUM LIKELIHOOD ESTIMATES FOR % WITHIN LIMITS
  WITH APPROXIMATE 95% CONFIDENCE LIMITS
```

	ESTIMATE $\hat{R}(500)$	LOWER LIMIT $\underset{\sim}{R}(500)$	UPPER LIMIT $\tilde{R}(500)$
PCT	99.46284%	89.66245	99.97471

Figure III.8A
STATPAC Output on Normal Fit to Wire Strength

* MAXIMUM LIKELIHOOD ESTIMATES FOR DIST. PARAMETERS
WITH APPROXIMATE 95% CONFIDENCE LIMITS

PARAMETERS	ESTIMATE	LOWER LIMIT	UPPER LIMIT
CENTER	$\hat{\mu}_B$= 1522.314	$\underset{\sim}{\mu}_B$= 1283.975	$\tilde{\mu}_B$= 1760.654
SPREAD	$\hat{\sigma}_B$= 434.9267	$\underset{\sim}{\sigma}_B$= 279.7427	$\tilde{\sigma}_B$= 676.1973

* COVARIANCE MATRIX

PARAMETERS	CENTER $\hat{\mu}_B$	$\hat{\sigma}_B$ SPREAD
CENTER $\hat{\mu}_B$	14787.01	
SPREAD $\hat{\sigma}_B$	4466.143	9589.641

PCTILES

* MAXIMUM LIKELIHOOD ESTIMATES FOR DIST. PCTILES
WITH APPROXIMATE 95% CONFIDENCE LIMITS

PCT. P	$\hat{y}_P$ ESTIMATE	$\underset{\sim}{y}_P$ LOWER LIMIT	$\tilde{y}_P$ UPPER LIMIT
0.1	178.1635	-371.9489	728.2759
0.5	401.8403	-59.57930	863.2599
1	510.3331	90.32051	930.3456
5	806.7678	490.4355	1123.100
10	964.8561	694.0329	1235.679
20	1156.342	923.9285	1388.756
50	1522.314	1283.975	1760.654
80	1888.286	1553.975	2222.598
90	2079.772	1678.140	2481.405
95	2237.861	1776.367	2699.354
99	2534.295	1954.560	3114.030

PERCENT(500. LIMIT)

* MAXIMUM LIKELIHOOD ESTIMATES FOR % WITHIN LIMITS
WITH APPROXIMATE 95% CONFIDENCE LIMITS

	ESTIMATE	LOWER LIMIT	UPPER LIMIT
	$\hat{R}(500)$	$\underset{\sim}{R}(500)$	$\tilde{R}(500)$
PCT	99.06270	88.25481	99.93277

Figure III.8B
STATPAC Output on Normal Fit to Bond Strength

The Life Distribution When All Failure Modes Act

When a product is in actual use, any mode can cause failure. The following paragraphs give the ML estimate of life distribution when all failure modes act. The estimate employs the product rule for reliability of a series system.

Suppose the product reliability at a given age is to be estimated. First obtain the ML estimate of the reliability for that age for each failure mode as described above. Then the ML estimate of product reliability is the product of the reliability estimates for each failure mode. Confidence limits for such a reliability are described by Nelson (1982, Chap. 8, Sec. 4). The estimate and confidence limits can be calculated for any number of ages as needed.

The connection reliability for 500 mg is estimated as follows. The ML estimates of reliability at 500 mg obtained from Figures III.8A and B are 0.995 for wires and 0.991 for bonds. So the ML estimate for connections is $0.995 \times 0.991 = 0.986$. The estimate of the fraction of connections with strength below 500 mg is $1 - 0.986 = 0.014$, a point on the curve in Figure III.7A. That curve is the ML estimate of the strength distribution of connections. It is obtained from repeating the above calculation for various strengths.

The Life Distribution after Certain Failure Modes Are Eliminated

For certain products, some failure modes can be eliminated by proposed design changes. This section shows how to use past data to estimate in advance the life distribution that would result.

Suppose the reliability of the new design at a given age is to be estimated. First obtain the ML estimate of the reliability at that age for each remaining failure mode, as described above. The ML estimate of the new design reliability for that age is the product of those reliability estimates.

Suppose bond failures could be eliminated, and the new connection reliability for 500 mg is to be estimated. The wire reliability estimate at 500 mg is 0.995. So the ML estimate of the reliability of improved connections is 0.995. The estimate of the percentage below 500 mg is 0.5%; this is marginal. So the wire also needs improvement.

7. INSPECTION DATA

For some products, a failure is found only on inspection, for example, a cracked component inside a turbine. There are two types of inspection data: 1) quantal-response data (exactly one inspection on each unit) and 2) interval data (any number of inspections on a unit). Hazard plotting and previous maximum likelihood (ML) methods apply to life data where each failure time is known exactly. Suitable analyses of inspection data take into account that a failure is known only to be in an interval. It is wrong to treat the inspection time when a failure is found as the failure time.

Graphical and maximum likelihood methods for such data are given in detail by Nelson (1982, Chapter 9). The following example illustrates ML fitting to quantal-response data by STATPAC of Nelson and others (1978).

Turbine Wheel Data

Table III.4 shows part of a large data set on turbine wheel age (in hours) at inspection and condition (− denotes cracked and + denotes not cracked).

An estimate of the distribution of time to crack initiation was needed to schedule regular inspections. Also, engineering wanted to know if the hazard (crack initiation) rate increases with wheel age; this would require replacement of wheels by some age when the risk of cracking gets too high.

STATPAC gives the ML estimates of percentiles, which plot as a straight line in Figure III.9. Their two-sided approximate 95% confidence limits are curves. Nelson (1982, Ch. 9) gives more details.

STATPAC output from the Weibull fit to the wheel data appears in Figure III.10. The β estimate is 2.09; this is greater than 1 and indicates that the hazard rate increases. Approximate 95% confidence limits 1.65 and 2.66 do not enclose 1 (constant failure rate). This is statistically significant (convincing) evidence that the hazard rate increases with age. So uncracked wheels should be replaced at some age.

Table III.4
Wheel Crack Initiation Data

```
3322+
4009+
1975-
1967-
1892-
2155+
2059+
4144+
1992+
1676+
4079+
2278-
1366+
etc.
```

+ unfailed on inspection
- found failed on inspection

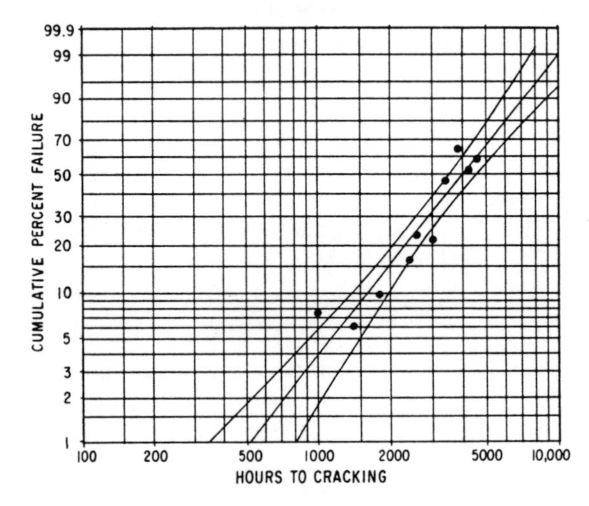

Figure III.9
Weibull Probability Plot of ML Fit to the Wheel Data

* MAXIMUM LIKELIHOOD ESTIMATES FOR DIST. PARAMETERS
WITH APPROXIMATE 95% CONFIDENCE LIMITS

PARAMETERS		ESTIMATE	LOWER LIMIT	UPPER LIMIT
CENTER	scale $\hat{\alpha}$	4809.557	4217.121	5485.219
SPREAD	shape $\hat{\beta}$	2.091212	1.645579	2.657525

* COVARIANCE MATRIX

PARAMETERS		CENTER $\hat{\alpha}$	SPREAD $\hat{\beta}$
CENTER	$\hat{\alpha}$	104048.0	
SPREAD	$\hat{\beta}$	-58.57977	0.6537992E-01

* CORRELATION MATRIX

PARAMETERS	CENTER	SPREAD
CENTER	1.0000000	
SPREAD	-0.7102457	1.000000

PCTILES

* MAXIMUM LIKELIHOOD ESTIMATES FOR DIST. PCTILES
WITH APPROXIMATE 95% CONFIDENCE LIMITS

PCT.	ESTIMATE	LOWER LIMIT	UPPER LIMIT
0.1	176.8618	87.44939	357.6937
0.5	382.2012	226.8173	644.0328
1	533.0458	342.0775	830.6241
5	1162.162	892.7043	1512.955
10	1639.675	1357.619	1980.331
20	2347.489	2079.196	2650.402
50	4036.353	3631.013	4486.942
80	6038.581	5071.816	7189.627
90	7166.606	5806.924	8844.656
95	8127.646	6407.249	10309.98
99	9983.057	7515.869	13260.13

Figure III.10
STATPAC Output for Weibull Fit

CHAPTER IV SURVEY OF OTHER TOPICS

This chapter briefly surveys some topics not covered above. For details, refer to books listed in Section 12, particularly Chapter 13 of Nelson (1982).

1. COMPARISONS WITH HYPOTHESIS TESTS

Often one wants to compare samples from different vendors, designs, lots of raw material, etc., to determine if they differ statistically significantly (convincingly). One can then make appropriate decisions. Such comparisons can be made subjectively with the aid of data plots or estimates and confidence intervals described above. More formal, objective comparisons may be made with hypothesis tests described in books listed in Section 12.

2. SYSTEM RELIABILITY MODELS AND THEIR ANALYSES

Reliability books (listed in Section 12) present methods for evaluating system reliability from the component reliabilities. Such methods help designers determine if a design meets reliability requirements. Also, such methods show how design reliability can be improved through better components and other system designs, for example, with redundant components. Such reliability analyses can be repeated for different definitions of system failure and operating environments. Major approaches to system reliability analyses are FMEA, fault tree and other analyses, coherent structures, common mode failures, and simulation.

3. COMPONENT FAILURE RATES

There are many handbooks on component failure rates. Best known are:

MILITARY HANDBOOK 217C Reliability Prediction of Electronic Equipment, 9 April 1979, available from Naval Publications and Forms Center, 5801 Tabor Ave., Philadelphia, PA 19120.

Government-Industry Data Exchange Program (GIDEP) Reliability-Maintainability Data Summaries, available from GIDEP Operations Center, Corona, CA 91720. This includes electronic and mechanical components.

4. MAINTAINABILITY

Repairable systems need to be designed for easy maintenance — fast and cheap repair. Statistical maintainability deals with distributions and data analyses for time to repair. Maintainability demonstration deals with sampling plans and data analyses to estimate the distribution of time to repair. For example, demonstration plans compare the mean and percentiles of the distribution of time to repair with specified values. Key references are:

Bird, G.T. (1969), "MIL-STD-471, Maintainability Demonstration," *J. Qual. Technol.* **1,** 134-148.

Gertsbakh, I.B. (1977), *Models of Preventive Maintenance*, North Holland, New York.

5. AVAILABILITY

Repairable systems are working (up) or failed and awaiting or undergoing repair (down). The long run fraction or the time that a system is up is called the *steady state availability.* Models and analyses for this availability are like those for reliability. There are some methods for analyzing failure and repair time data to get availability information. Key references are:

Barlow, R.E., and Proschan, F. (1965), *Mathematical Theory of Reliability,* Wiley, New York.

Sandler, G.H. (1963), *System Reliability Engineering*, Prentice-Hall, Englewood Cliffs, NJ.

6. ESTIMATION OF SYSTEM RELIABILITY FROM COMPONENT DATA

One may want an estimate and confidence limits for system reliability from component reliability data. This topic is surveyed by Mann, Schafer, and Singpurwalla (1974, Chap. 10).

7. BAYESIAN METHODS

Bayesian methods provide a formal means of including subjective information on product or component reliability in a reliability analysis of a system (or component). Such methods are used for probabilistic analysis of system reliability and for data analysis. Bayesian methods are surveyed by Mann, Schafer, and Singpurwalla (1974, Chap. 8). Locks (1973, Chap. 7) gives a simple introduction. Martz and Waller (1982) devote a book to it.

8. RELIABILITY DEMONSTRATION AND ACCEPTANCE TESTS

Reliability demonstration tests are used to assess whether product performance meets a requirement that a parameter of the life distribution surpass a specified value. For example, the mean life of an exponential distribution must exceed a specified value. Test plans specify the sample size, test time, and criteria for the product passing or failing the test. Many plans have been given in standards, books, and the literature dealing with reliability and quality control. Pabst (1975) and Schilling (1982) review a number of military standards.

9. RELIABILITY GROWTH (DUANE PLOTS)

In many development programs, hardware reliability grows during design, development, testing, and actual use — due to *continuing* engineering improvements in design, manufacture, and operation. Reliability managers employ predictions of hardware reliability that will result from continued engineering improvements. Reliability growth models and data analyses provide such predictions, confidence limits for the true reliability (or failure rate), and prediction limits for future numbers of failures. Key references are:

Codier, E. O. (1968), "Reliability Growth in Real Life," *Proceedings of the 1968 Annual Symposium on Reliability*, pp. 458-469.

Crow, L. H. (1977), "Confidence Interval Procedures for Reliability Growth Analysis," U.S. Army Material Systems Analysis Activity, Technical Report No. 197, Aberdeen Proving Ground, MD 21005. Also available as document AD-A044788 from Defense Documentation Center, Defense Logistics Agency, Cameron Station, Alexandria, VA 22314.

MIL-HDBK-189, "Reliability Growth Management," Department of Defense, Washington, DC (1981). Available from the Commanding Officer, Naval Publications and Forms Center, 5801 Tabor Ave., Philadelphia, PA 19120.

10. MODELS AND DATA ANALYSES FOR REPAIRABLE PRODUCTS

Failed components of many systems and products are replaced or repaired. One often wants to use field or test data to predict the expected number of repairs as a function of calendar time. Similar models and data analyses are used for burn-in applications. Duane plotting is a special application. See Goldberg (1981) listed in Section 12.

11. LIFE AS A FUNCTION OF OTHER VARIABLES (ACCELERATED TESTING)

Product life may depend on variables arising in design, manufacture, operation, etc. There are regression and analysis-of-variance models and analyses for such data; see Lawless (1982), Elandt-Johnson and Johnson (1980), Gross and Clark (1975), Kalbfleisch and Prentice (1980), and Lee (1980) listed in Section 12.

A special case, accelerated life testing of products and materials yields life information quickly and cheaply. Endurance testing of dielectrics and fatigue testing of metals are important applications. Test units are overstressed, resulting in shorter lives than under normal conditions. A model fitted to the accelerated data is used to estimate product life under normal conditions. Nelson (1974) surveys accelerated testing models and data analyses, and Meeker (1979) presents a large bibliography including applications.

12. RECENT BOOKS WITH RELIABILITY AND LIFE DATA ANALYSIS

Books below present reliability models and data analyses. The annotated list omits books on reliability management, physics of failure, and other nonmathematical methods. Early books lack important new developments and are omitted. These books contain further information on the topics surveyed above.

Bain, L. J. (1978), *Statistical Analysis of Reliability and Life-Testing Models*, Dekker, New York. Advanced mathematical development with engineering applications.

Barlow, R. E. and Proschan, F. (1975), *Statistical Theory of Reliability and Life Testing*, Holt, Rinehart, and Winston, New York. Advanced mathematical development.

Bury, K.V. (1975), *Statistical Models in Applied Science*, Wiley, New York. Intermediate selected topics.

Elandt-Johnson, R. C. and Johnson, N. L. (1980), *Survival Models and Data Analysis*, Wiley, New York. Advanced mathematical development with biomedical applications.

Fitch, W. C., Wolf, F. K., and Bissinger, B. H. (1982), *The Estimation of Depreciation*, to be published. Applied intermediate presentation.

Gertsbakh, I. B. (1977), *Models of Preventive Maintenance*, North-Holland Amsterdam.

Goldberg, H. (1981), *Extending the Limits of Reliability Theory*, Wiley, New York. Intermediate mathematical development of renewal theory for repairable systems.

Gross, A. J. and Clark, V. A. (1975), *Survival Distributions: Reliability Applications in the Medical Sciences*, Wiley, New York. Intermediate with biomedical applications.

Halpern, S. (1978), *The Assurance Sciences: An Introduction to Quality Control and Reliability*, Prentice-Hall, Englewood Cliffs, NJ. Introduction to selected topics.

Kalbfleisch, J. D. and Prentice, R. L. (1980), *The Statistical Analysis of Failure Time Data*, Wiley, New York. Advanced with biomedical applications.

Kapur, K. C. and Lamberson, L. R. (1977), *Reliability in Engineering Design*, Wiley, New York.

Kaufman, A., Grouchko, D., and Cruon, R. (1977), *Mathematical Models for the Study of Reliability of Systems*, Academic, New York.

Lawless, J. F. (1982), *Statistical Models and Methods for Lifetime Data*, Wiley, New York. Advanced mathematical development with biomedical and engineering applications.

Lee, E. (1980), *Statistical Methods for Survival Data Analysis*, Lifetime Learning, Belmont, CA. Intermediate with biomedical applications.

Little, R. E. and Jebe, E. H. (1975), *Statistical Design of Fatigue Experiments*, Halstead, New York. Intermediate with metal fatigue applications.

Lloyd, D. K. and Lipow, M. (1977), *Reliability: Management, Methods and Mathematics*, 2nd ed., McGraw-Hill, New York. Intermediate, few recent data analysis methods.

Locks, M. O. (1973), *Reliability, Maintainability, and Availability Assessment*, Hayden, Rochelle Park, NJ. Introduction to engineering applications.

Mann, N.R., Schafer, R. E., and Singpurwalla, N. D. (1974), *Methods for Statistical Analysis of Reliability and Life Data*, Wiley, New York. Advanced mathematical development with engineering applications.

Martz, H. F. and Waller, R. A. (1982), *Bayesian Reliability Analysis*, Wiley, New York. Intermediate mathematical development.

Miller, R. (1981), *Survival Analysis*, Wiley, New York. Intermediate with biomedical applications.

Nelson, W. (1982), *Applied Life Data Analysis*, Wiley, New York. Intermediate with engineering applications.

Sinha, S. K. and Kale, B. K. (1980), *Life Testing and Reliability Estimation*, Halstead, New York. Intermediate mathematical development with engineering applications.

REFERENCES

General Electric reports and reprints marked with an asterisk * are available from the Technical Information Exchange, 5-321, General Electric Company Corporate Research & Development, Schenectady, NY 12345.

Bain, L. J. (1978), *Statistical Analysis of Reliability and Life-Testing Models: Theory and Methods,* Marcel Dekker, New York.

Block, H. W. and Savits, T. H. (1981), "Multivariate Distributions in Reliability Theory and Life Testing," Technical Report No. 81-13, Inst. for Statistics and Applications, Dept. of Math. and Statistics, Univ. of Pittsburgh, Pittsburgh, PA 15260.

David, H. A. and Moeschberger, M. L. (1979), *The Theory of Competing Risks,* Griffin's Statistical Monograph No. 39, Methuen, London.

General Electric Information Services Company (1979), "STATSYSTEM Users Guide," GE Information Services Company Publication 5707.12. Available from your GEISCO service representative or (800) 638-8730.

Glasser, M. (1965), "Regression Analysis with Censored Data," *Biometrics 21,* 300-307.

Grant, E. L. and Leavenworth, R. S. (1980), *Statistical Quality Control,* 5th ed., McGraw-Hill, New York.

Hahn, G. J. and Miller, J. M. (1968), "Methods and Computer Program for Estimating Parameters in a Regression Model from Censored Data," General Electric Research & Development Center TIS Report 68-C-277.*

Hahn, G. J. and Shapiro, S. S. (1967), *Statistical Models in Engineering,* Wiley, New York.

IMSL (1975), "Library 2 Reference Manual," 5th ed. (November 1975), International Mathematical and Statistical Libraries, Inc., 6th Floor, GNB Building, 7500 Bellaire Boulevard, Houston, TX 77036.

Jensen, F. and Petersen, N. E. (1982), *Burn-in: An Engineering Approach to the Design and Analysis of Burn-in Procedures,* Wiley, New York.

Lawless, J. F. (1978), "Confidence Interval Estimation for the Weibull and Extreme Value Distributions," *Technometrics 20,* 355-368.

Lieberman, G. and Owen, D. B. (1961), *Tables of the Hypergeometric Probability Distribution,* Stanford Univ. Press, Stanford, Calif.

Locks, M. O. (1973), *Reliability, Maintainability, and Availability Assessment,* Hayden, Rochelle Park, NJ.

Mann, N. R., Schafer, R. E., and Singpurwalla, N.D. (1974), *Methods for Statistical Analysis of Reliability and Life Data,* Wiley, New York.

Martz, H. F. and Waller, R. A. (1982), *Bayesian Reliability Analysis,* Wiley, New York.

McCool, J. I. (1974), "Inferential Techniques for Weibull Populations," Aerospace Research Laboratories Report ARL TR 74-0180, available from National Technical Information Services Clearinghouse, Springfield, VA 22151, publication AD A 009 645.

Meeker, W. Q. (1979), "Bibliography on Accelerated Testing," Dept. of Statistics, Iowa State Univ., Ames, Iowa 50011.

Meeker, W. Q. and Duke, S. (1979), "CENSOR — A User-Oriented Computer Program for Life Data Analysis," Department of Statistics, Iowa State University, Ames, Iowa 50011.

Nelson, Wayne (1970), "Confidence Intervals for the Ratio of Two Poisson Means and Poisson Predictor Intervals," *IEEE Trans. on Reliability R-19,* 42-49.*

Nelson, W. (1974), "A Survey of Methods for Planning and Analyzing Accelerated Life Tests," *IEEE Trans. on Electrical Insulation EI-9,* 12-18.*

Nelson, W. (1979), "How to Analyze Data with Simple Plots," Volume 1 of *The ASQC Basic References in Quality Control: Statistical Techniques*, J. Dudewicz, Editor. For sale from the Amer. Soc. for Quality Control, 230 W. Wells St., Milwaukee, WI 53203.

Nelson, Wayne (1982), *Applied Life Data Analysis,* Wiley, New York.

Nelson, W. B., Morgan, C. B., and Caporal, P. (1978), "1979 STATPAC Simplified — A Short Introduction to How to Run STATPAC, a General Statistical Package for Data Analysis," General Electric Company Corporate Research & Development TIS Report 78CRD276.* Outside General Electric, STATPAC may be obtained on license through Mr. M. Keith Burk, Technology Marketing Operation, GE Corporate Research & Development, 120 Erie Blvd., Schenectady, NY 12305. Phone: (518) 385-3801.

Pabst, W. R. Jr., editor (1975), *Standards and Specifications,* Publication 103, Amer. Soc. for Quality Control, 230 W. Wells St., Milwaukee, WI 53203.

Preston, D. L. and Clarkson, D. B. (1980), "SURVREG: An Interactive Program for Regression Analysis with Censored Survival Data," presented at the annual meeting of the American Statistical Association, Houston, TX.

Schilling, E. G. (1982), *Acceptance Sampling in Quality Control,* Marcel Dekker, New York.

Schmee, J., Gladstein, D., and Nelson, W. (1982), "Exact Confidence Limits for Parameters of the (Log) Normal Distribution from Maximum Likelihood Estimates and Singly Censored Samples," General Electric Co. Corp. Research & Development TIS Report 82CRD244* to appear in *Technometrics 27* (May 1985).

Appendix A. Chi-Square Percentiles χ^2 $(P;\nu)$

ν \ P	0.005	0.010	0.025	0.050	0.100	0.250	0.500
1	0.00004	0.00016	0.00098	0.00393	0.01579	0.1015	0.4549
2	0.0100	0.0201	0.0506	0.1026	0.2107	0.5754	1.386
3	0.0717	0.1148	0.2158	0.3518	0.5844	1.213	2.366
4	0.2070	0.2971	0.4844	0.7107	1.064	1.923	3.357
5	0.4177	0.5543	0.8312	1.145	1.610	2.675	4.351
6	0.6757	0.8721	1.2373	1.635	2.204	3.455	5.348
7	0.9893	1.239	1.690	2.167	2.833	4.255	6.346
8	1.344	1.646	2.180	2.733	3.490	5.071	7.344
9	1.735	2.088	2.700	3.325	4.168	5.899	8.343
10	2.156	2.558	3.247	3.940	4.865	6.737	9.342
11	2.603	3.053	3.816	4.575	5.578	7.584	10.34
12	3.074	3.571	4.404	5.226	6.304	8.438	11.34
13	3.565	4.107	5.009	5.892	7.041	9.299	12.34
14	4.075	4.660	5.629	6.571	7.790	10.17	13.34
15	4.601	5.229	6.262	7.261	8.547	11.04	14.34
16	5.142	5.812	6.908	7.962	9.312	11.91	15.34
17	5.697	6.408	7.564	8.672	10.09	12.79	16.34
18	6.265	7.015	8.231	9.390	10.86	13.68	17.34
19	6.844	7.633	8.907	10.12	11.65	14.56	18.34
20	7.434	8.260	9.591	10.85	12.44	15.45	19.34
21	8.034	8.897	10.28	11.59	13.24	16.34	20.34
22	8.643	9.542	10.98	12.34	14.04	17.24	21.34
23	9.260	10.20	11.69	13.09	14.85	18.14	22.34
24	9.886	10.86	12.40	13.85	15.66	19.04	23.34
25	10.52	11.52	13.12	14.61	16.47	19.94	24.34
26	11.16	12.20	13.84	15.38	17.29	20.84	25.34
27	11.81	12.88	14.57	16.15	18.11	21.75	26.34
28	12.46	13.56	15.31	16.93	18.94	22.66	27.34
29	13.12	14.26	16.05	17.71	19.77	23.57	28.34
30	13.79	14.95	16.79	18.49	20.60	24.48	29.34
40	20.71	22.16	24.43	26.51	29.05	33.66	39.34
50	27.99	29.71	32.36	34.76	37.69	42.94	49.33
60	35.53	37.48	40.48	43.19	46.46	52.29	59.33
70	43.28	45.44	48.76	51.74	55.33	61.70	69.33
80	51.17	53.54	57.15	60.39	64.28	71.14	79.33
90	59.20	61.75	65.65	69.13	73.29	80.62	89.33
100	67.33	70.06	74.22	77.93	82.36	90.13	99.33

From N. L. Johnson and F. C. Leone, *Statistics and Experimental Design in Engineering and the Physical Sciences*, 2nd ed., Wiley, New York, 1977, Vol. 1, pp. 511-512. Reproduced by permission of the publisher and the Biometrika Trustees.

Appendix A. Chi-Square Percentiles χ^2 $(P;v)$ (*Continued*)

v \ P	0.750	0.900	0.950	0.975	0.990	0.995	0.999
1	1.323	2.706	3.841	5.024	6.635	7.879	10.83
2	2.773	4.605	5.991	7.378	9.210	10.60	13.82
3	4.108	6.251	7.815	9.348	11.34	12.84	16.27
4	5.385	7.779	9.488	11.14	13.28	14.86	18.47
5	6.626	9.236	11.07	12.83	15.09	16.75	20.52
6	7.841	10.64	12.59	14.45	16.81	18.55	22.46
7	9.037	12.02	14.07	16.01	18.48	20.28	24.32
8	10.22	13.36	15.51	17.53	20.09	21.96	26.12
9	11.39	14.68	16.92	19.02	21.67	23.59	27.88
10	12.55	15.99	18.31	20.48	23.21	25.19	29.59
11	13.70	17.28	19.68	21.92	24.72	26.76	31.26
12	14.85	18.55	21.03	23.34	26.22	28.30	32.91
13	15.98	19.81	22.36	24.74	27.69	29.82	34.53
14	17.12	21.06	23.68	26.12	29.14	31.32	36.12
15	18.25	22.31	25.00	27.49	30.58	32.80	37.70
16	19.37	23.54	26.30	28.85	32.00	34.27	39.25
17	20.49	24.77	27.59	30.19	33.41	35.72	40.79
18	21.60	25.99	28.87	31.53	34.81	37.16	42.31
19	22.72	27.20	30.14	32.85	36.19	38.58	43.82
20	23.83	28.41	31.41	34.17	37.57	40.00	45.32
21	24.93	29.62	32.67	35.48	38.93	41.40	46.80
22	26.04	30.81	33.92	36.78	40.29	42.80	48.27
23	27.14	32.01	35.17	38.08	41.64	44.18	49.73
24	28.24	33.20	36.42	39.36	42.98	45.56	51.18
25	29.34	34.38	37.65	40.65	44.31	46.93	52.62
26	30.43	35.56	38.89	41.92	45.64	48.29	54.05
27	31.53	36.74	40.11	43.19	46.96	49.64	55.48
28	32.62	37.92	41.34	44.46	48.28	50.99	56.89
29	33.71	39.09	42.56	45.72	49.59	52.34	58.30
30	34.80	40.26	43.77	46.98	50.89	53.67	59.70
40	45.62	51.80	55.76	59.34	63.69	66.77	73.40
50	56.33	63.17	67.50	71.42	76.15	79.49	86.66
60	66.98	74.40	79.08	83.30	88.38	91.95	99.61
70	77.58	85.53	90.53	95.02	100.4	104.2	112.3
80	88.13	96.58	101.9	106.6	112.3	116.3	124.8
90	98.65	107.6	113.1	118.1	124.1	128.3	137.2
100	109.1	118.5	124.3	129.6	135.8	140.2	149.4